高等院校土建学科双语教材（中英文对照）

◆ 建筑学专业 ◆

BASICS

空 间 设 计
SPATIAL DESIGN

[德] 欧利奇·埃克斯纳（Ulrich Exner）
迪特里希·普雷塞尔（Dietrich Pressel） 编著

董 慰 张 宇 译

中国建筑工业出版社

著作权合同登记图字：01-2009-7702 号

图书在版编目（CIP）数据

空间设计/（德）埃克斯纳，（德）普雷塞尔编著；董慰，张宇译. —北京：中国建筑工业出版社，2014.10
高等院校土建学科双语教材（中英文对照）◆ 建筑学专业 ◆
ISBN 978-7-112-17029-6

Ⅰ.①空…　Ⅱ.①埃…②普…③董…④张…　Ⅲ.①室内装饰设计-双语教学-高等学校-教材-汉、英　Ⅳ.①TU238

中国版本图书馆 CIP 数据核字（2014）第 140612 号

Basics：Spatial Design/Ulrich Exner，Dietrich Pressel
Copyright © 2009 Birkhäuser Verlag AG，P.O. Box 133，4010 Basel，Switzerland
Chinese Translation Copyright © 2014 China Architecture & Building Press
All rights reserved.
本书经 Birkhäuser Verlag AG 出版社授权我社翻译出版

责任编辑：孙书妍　　责任设计：陈　旭　　责任校对：陈晶晶　刘　钰

高等院校土建学科双语教材（中英文对照）
◆ 建筑学专业 ◆

空间设计

［德］欧利奇·埃克斯纳（Ulrich Exner）
　　　迪特里希·普雷塞尔（Dietrich Pressel）　　编著
董　慰　张　宇　　　　　　　　　　　　　　　译
*
中国建筑工业出版社出版、发行(北京海淀三里河路9号)
各地新华书店、建筑书店经销
北京嘉泰利德公司制版
北京建筑工业印刷厂印刷
*
开本：880×1230 毫米　1/32　印张：5　字数：205 千字
2014 年 9 月第一版　　2019 年 11 月第三次印刷
定价：**27.00** 元
ISBN 978-7-112-17029-6
　　　　　（32400）

中文部分目录

CONTENTS

序

　　空间规划和设计是建筑设计的基本内容。不管是景观、城市空间还是室内空间，都可以将一些可以用来比较的原则和参量应用到空间设计与感知中。建筑师、城市规划师或其他个人都可以有意识地设计空间，但空间使用状况与时间印迹也能对空间形式及其改变产生相当重要的影响。此外，空间的感知和评价并不客观，往往依赖于使用者的个人感知及其社会文化背景。因此，我们提出一系列宽泛、可能的设计方法和丰富、多样的设计技术。

　　本书通过对空间现象的审视来关注现实的核心内容，这些空间现象与具体的空间功能和专业学科并不相关。本书也讨论了人类用于感知环境的不同感官感觉、如何获得和传递感官刺激的方式，以及我们如何基于个人经验对其进行评估。为了更好地解释不同的空间，本书对不同空间类型及其各自特征进行了情境化的描述。"空间设计的参量"一章中讨论了所有类型空间设计必要的基本原则，并在"空间设计的要素和方法"一章中，通过对每个设计方法及其实际案例的阐述对上述基本原则加以证明。作者表达了他对空间特征的深刻理解，提出了可供设计师使用的空间设计方法。

编辑：伯特·比勒费尔德（Bert Bielefeld）

FOREWORD

Planning and designing spaces is an essential aspect of architectural design. Comparable principles and parameters can be identified, irrespective of landscapes, urban spaces, or spaces inside buildings, and applied to design and perceive space. Space can be designed consciously by architects, urban planners, or other individuals, yet use and the passing of time have an equally important effect on its shape and transformation. Moreover, there is no objective perception or evaluation of space; it is always conditional upon the viewer or user's individual senses and socio-cultural background. This allows a wide spectrum of possible approaches and a wealth of design options.

Basics Spatial Design is a continuation of our thematic series on design. It focuses on one of the practice's key components by examining the phenomenon of space, independently of specific functions and professional disciplines. As an important introduction, the book also discusses the different senses humans employ to perceive their environment, how people process these sensory stimuli, and how we base our evaluations on personal experience. To explain the range of different spaces, the book goes on to describe and contextualize the different spatial typologies and their respective characteristics. Basic principles that are essential to designing all types of spaces are discussed in the chapter "The parameters of spatial design" and then substantiated in the chapter "Elements and Means of Spatial Design," using individual design methods and actual examples. The authors hope to convey a deeper understanding of the specific attributes of spaces, and the ways in which designers can consciously influence their subsequent effect.

Bert Bielefeld, Editor

INTRODUCTION

Space is fundamental to human existence, and much of the spatial environment is designed by people. Day-to-day life always takes place within a space, whether it is a landscape, a city, a house, or a room. People naturally trust that their built or natural environment is permanent, despite the fact that earthquakes or war can suddenly destroy that very environment. People perceive space with their senses directly, individually, and always in a new and fresh way. There are spaces in which we enjoy or do not enjoy walking, relaxing, dreaming, or working. A forest or a street might seem inviting during the day but threatening at night. Within seconds, a spatial situation can feel too close or too large, safe or threatening, inviting or repulsive, which are all impressions that influence our behavior accordingly. Hikers always deliberately choose a resting place according to particular criteria: the sun is shining, the wind is not too strong, it is sufficiently cool, has a pleasant view, and sounds from the environment are absorbed well enough so as not to disturb the sought-after tranquility. The atmosphere of a place such as this is difficult to describe in detail, because various aspects come together simultaneously to make an impression; they are not perceived and analyzed individually.

People design their spatial environment according to their needs for protection against the forces of nature, their various behavioral patterns, work and life style, and their desires and philosophies. Yet a large part of the spatial environment is determined by others or is pre-given, often by the private interests of others, according to natural factors, or the will of the political majority. Constructed spaces can stimulate the senses and the mind through form, materiality, and light or color. Their dimensions can provide either shelter or security, and their design can generate feelings of surprise, astonishment, joy, or wellbeing. Inventing a spatial container is at the same time the invention of a way to enliven it. Spatial design, as a built implementation, can also be described as the cultural-ideological, site-specific, economical, political, social or use-determining parameters that define human existence. These parameters are subject to constant change and always influence constructed spaces. In spatial design, requirements and concepts should be recognized that will be applicable to an individual or relevant to a group, for a millennium, or perhaps only for a few hours.

Spatial design can be generally defined as any type of active spatial appropriation, whether it is a room or a landscape. At the center of this group is space as a relationship, perceived sensorially and cognitively, between things, bodies, or elements of the activated nature. Below,

Fig.1:
Cans provide the greatest amount of volume in relation to the most
amount of external material, plus a choice of surface design.

we discuss human perception of the built and natural environment, the
characteristic phenomena of space, and the means and elements available for designing it.

SPATIAL PERCEPTION

The prerequisite for any spatial design and its effect is the human sensory and cognitive perception of the surrounding environment. All of the sensory stimuli conveyed by the space are processed by the brain, which influences how an individual feels, behaves, and moves.

Humans are believed to possess up to thirteen senses, including the five main senses of sight, hearing, touch, smell, and taste, as well as balance. Some people do not have access to all of the senses, or are not able to perceive or fully perceive certain sensory stimuli such as light or sound. The sense of equilibrioception is responsible for perceiving gravity, and therefore spatial verticality, as the constant orientation in space.

Spatial perception serves our individual, day-to-day basic orientation, without our needing to absorb all of the spatial characteristics completely. We are constantly using new spaces in our daily lives. Much of a space's information is processed so quickly by the senses and the cognitive system that it automatically influences our behavior without the need to first activate our thinking process. The human processing of perception and information quickly allows a space to appear cozy or uncomfortable, claustrophobic or protective, without perceiving the spatial characteristics individually. We know the moment we enter a café whether we like the atmosphere or not.

Spatial perception is individual. After a long period of time, adults see the place where they spent their childhood as small, although they remember it as being large. At the same time, there are many spatial characteristics that several people perceive in a similar way. Orientation systems, for one, would not function otherwise. Perceiving the spatial environment mostly occurs while we are in motion, which can be encouraged by a space's particular attributes.

\\ Note:
The cognitive system is the term used for the human function associated with perception, learning, remembering, and thinking; in other words, human thoughts and mental processes.

\\ Note:
Sensory perception is aesthesis in Greek. In philosophy, the term "aesthetics" is used to describe the theory of sensory perception. In everyday speech however, aesthetic is now used as a synonym for beautiful.

Fig.2:
The specific attributes of material are perceived with the close senses

Fig.3:
Directed view, restricted to the human field of vision

Fig.4:
Estimating the distance to the Himmels-treppe (Sky Stairs) in the Moroccan desert

Fig.5:
Arriving at the Himmelstreppe after walking for two hours

CLOSE AND DISTANCE SENSES

Perception is mainly a product of the five human senses of sight, hearing, touch, taste, and smell. The intensity of these senses varies from person to person. > Tab. 1 They can only produce a complete sensory impression by working together—for example when a picture of the rough surface of a wooden board evokes the impression of how the depth of its grain might feel and how it would smell.

Close senses

Direct contact with the perceivable object is created by the close senses of smell, touch, and taste. None of these three requires light, and they are for the most part constantly available. The sense of touch is essential to a feeling of wellbeing in a space, because contact with the spatial shell is made through the skin.

> 🔎

Tab.1: Intake capacity of the five main senses in bits per second				
Sight	**Touch**	**Hearing**	**Smell**	**Taste**
10,000,000	1,000,000	100,000	100,000	1,000

Tab.2: Visibility range in different atmospheric conditions					
Very clear visibility	**Clear visibility**	**Slightly cloudy visibility**	**Cloudy visibility**	**Very cloudy, light fog**	**Snow flurries, thick fog**
50-80 km	20-50 km	10-20 km	4-5 km	2 km	0.01 km

Distance senses   Acoustic and visual signals also work reciprocally while we are in the process of perceiving. Neural connections structure these signals and provide information about orientation in the direct human environment. Hence, visual perception is more selective when accompanied by specific acoustic signals than in an acoustically diffuse space. Through the eye's lens, visual signals project a two-dimensional image of the environment onto the retina. With the help of the neuronal structure of the brain and one's own personal experience, this image can be perceived as a spatially complex composite. Interpreting visual signals is conditional upon individual experience.

As is evident in the mountain range in the background of Figs. 4 and 5, contours are difficult to perceive at a distance and have a flat appearance. This makes it hard to estimate distance. By contrast, the atmospheric and visibility conditions › Tab. 2 that are familiar in Central Europe allow a relatively reliable estimation of distance in a typical landscape.

ρ
\\ Example:
The qualities of a material are touched, smelt, and seen. They will be evaluated as pleasant if all three individual factors are experienced as being in balance (see Fig. 2).

THE COGNITIVE SYSTEM

As described above, sensory impressions of space are more or less consciously interpreted by the intellect or the cognitive system, and influence our behavior, thoughts, and emotions. A spatial element can generate an instinctive behavior, be perceived as a signifier, or trigger memories.

This type of spatial perception is similar to reading a text. Analogously to the theories and methods of linguistics, sensorially perceived stimuli are "read" from spatial elements as signs, and their meaning is processed and interpreted by the human intellect. The elements of a space are hence seen as data transmitters that relay more about the elements than merely conveying their direct presence.

PHENOMENOLOGY OF SPACE

The philosophy of phenomenology represents the theory that spatial experience is directly influenced by human perception, which means that human behavior in the world is defined by sensory perception. Sensation and awareness were already assigned to the body before the thinking process was added to them. In the course of human development, physical experience has molded people's ideas about things, space, and time. Since human existence and the body are inseparably related to space, spatial design is significant in terms of learning as well as the general acquisition of knowledge.

Fig.6:
Impressive spatial atmosphere in a mosque in Istanbul

14

The atmospheric effect of a space is essential to human well-being—yet it is difficult to precisely define or to gauge, and can only be partially justified using analytical methods. Its diffuse qualities make it difficult to plan, present, or understand. A room lit by candles is generally considered "cozy." Yet the flickering flame of the candle, the colorful glow, and the diffuse darkness of the spatial borders' surfaces are not the only reasons behind this atmospheric spatial impression. In addition to these visual aspects, other sensory stimuli such as the scent of wax, the warmth of the flame, and its occasional, quiet sizzling sound all account for the inviting atmosphere. › Fig. 6

TYPES OF SPACES

Many different spatial forms are influenced by the same uses, human ideas, behavioral patterns and needs, or comparable site-specific conditions. They form spatial archetypes that can be found in different cultures depending on climate, region, and point in time. Therefore, uses such as residential, production, or the practice of religion can often be read in architectural form, meaning that the spatial shell and structural design clearly indicate the actions that take place within. › Fig. 7

The next section will introduce some common types of spaces, the uses of which can easily be read in the spatial design. But use alone does not determine form; other relevant parameters of spatial design will be explained in the section below.

Although people are constantly altering and adapting spaces to suit their changing needs, many of the structural features of a space remain consistent.

FUNCTIONAL SPACES

Spatial forms are always influenced by their functions. Every constructed building and space is a site for human interaction, dealings, rituals, games, and spectacles. These actions determine the spatial design to a great extent and, in turn, the spatial features influence the user and the functions. A space may be the necessary container for a certain action or may not have a specific function assigned to it at all. Spatial

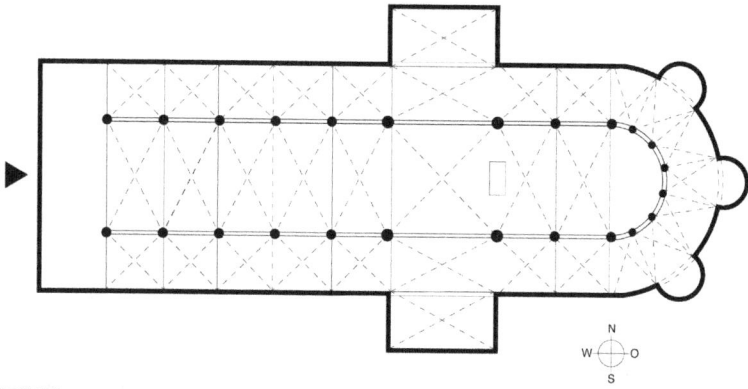

types can be identified and distinguished by whether or how clearly a specific function can be recognized in their structural design. Specific architectural requirements can strongly influence a spatial design and, if this composite has been built several times, create an architectural category. Infrastructure and engineering buildings are often very directly based on a specific use, making other, subsequent uses almost impossible. › Fig. 9

The opposite of this example is a spatial type that is open to several uses, a functional ability that also influences its spatial design. Hence, a public city square has only a few distinct assigned uses. Its size alone allows different activities that include individuals casually spending time, demonstrations, summer festivals, and weekly markets.

\\ Example:
A basilica is a precise spatial form that has been built throughout history in many different variations (it was originally a secular construction). The elongated hall is entered from the west. It is aligned with the apse, which mostly faces east in the direction of Jerusalem. The altar for performing religious rituals was situated here in clear public view (see Fig. 8).

Fig.9:
Technical requirements create spatial categories, such as the historic wind towers that have been traditionally used for centuries in Arab countries for ventilation, or the equally familiar cooling towers of power stations in industrial countries.

GENIUS LOCI

Site-specific
spatial type

The attributes of a specific site create the spatial type because they have a fundamental influence on the spatial structure and form. An overhanging precipice requires a different house and supporting frame than a flat piece of ground.

In addition to these factors, local wind, temperature, and lighting conditions also influence the choice of spatial alignment, the type and number of openings, or the specific attributes of the spatial shell. › Fig. 10

Fig.10:
A spatial type determined by local conditions (cave dwellings)

Fig.11:
Residential or office containers are largely not site-specific and can be used almost anywhere.

In comparison to this, there are many non-site specific spatial forms such as an airport terminal, which are universally applicable and form different contextual or functional references. Even industrially manufactured residential or office containers are relatively non-site-specific and are designed accordingly. › Fig. 11

Dimension

The specific scale of a landscape, a city, a street, or a room makes them spatial types that determines the activities they can accommodate as well as the significance attributed to them. A room can hold only a certain number of objects, people, and activities, and is perceived as a space with a private or semi-public character for a small number of people. A city square, on the other hand, is a suitably large space that can accommodate many people's day-to-day activities, such as work, shopping, eating, living, and communicating. Nest, territory, and universe represent three of the spatial dimensions that define human existence: private space, familiar environment, and public area.

Material

Material is an essential factor of a site-specific architectural type. A construction material that is available in large amounts locally creates typical spatial structures that can be found again and again, and their forms can be traced to the particular attributes and availability of the material. › Fig. 12

ρ

\\ Example:
If the development site is on a street above an incline, the access level can be placed at the upper floor, and the ground plan and spatial form can be adjusted accordingly.

Fig.12:
Brickearth houses in Southeast Asia made from local materials and a timber frame building in a densely wooded area

PRIVATE AND PUBLIC

Spaces are characterized by their level of public accessibility. Depending on their assigned uses, dimensions, and qualities, spaces can have either a private or public nature, which is quickly recognized and will directly influence our behavior while in the space. The borders between private and public are often blurred because private and public uses often intermix or change. The private or public character of a space is defined by dimensions, the degree of social control, and permeability, meaning the number and types of openings in its spatial shell.

Public spaces

A public space includes all accessible, open spaces within the built structure or community, and always exists where the general public uses space in different ways. It is simultaneously a space for movement, activities, information, and lingering. This is where groups and individuals from different social classes, nationalities, and cultures can meet and communicate. They can conduct business, express their opinions, and gather ideas directly without having to experience them through the media. Public space is shaped by its dimensions. It generally has to provide enough space for people to comfortably move about, yet cars, streets, and trains also determine its dimensions because public space is also a primary public traffic and transportation circuit. Sequences of movement and activities can be directed and controlled by designing this type of space, and it consequently also has political significance. Its design is often reshaped and transformed over the course of time and is witness to a multitude of uses and meanings.

Social control

Public space offers greater freedom of movement than the comparably small, private space. Social control and surveillance by others restricts and protects activities in public space, because it helps to maintain socially accepted standards of behavior. > Fig. 13

> 📎

Fig.13:
Urban public space

Fig.14:
Public square in Valencia

A lack of social control can quickly turn a space into something inhospitable; public activities no longer take place and there are no attractive amenities that encourage a visitor to linger.

Squares

Public squares and buildings are frequently assigned symbolic or prestigious functions that influence the development of urban structures in regards context. Political, scientific, economic, or religious developments, as well as new means of transportation and communication, are constantly changing the design, significance, and use of public space. Public space is often a spatial type that has been influenced by the site in question. Special smells, a certain sound absorption, climatic conditions, or people's site-specific clothing, movements, and activities form the overall impression and determine the use of the space. Because of differences in culture and climate, public spaces and the way in which they are designed or animated differ fundamentally between southern and northern countries. › Fig. 14

\\ Note:
People in a busy and hence socially controlled public space would normally intervene or assist if a violent attack takes place. In a public landscape space, this level of control is missing, which produces a feeling of almost unlimited freedom that can also turn into fear.

Fig.15:
A line of houses with billboards and three-dimensional objects

Wherever people can be found, public spaces will also be defined by private interests or the will of the political majority. Control over the use and design of public space is an expression of power.

Public space is often shaped by the need for good orientation and the great number of signs and elements it requires.

Private space is a spatial type that is meant to protect the privacy of the individual. This is a place for activities that are not observed by the general public.

> Private space

\\ Note:
Private economic interests have turned facades in public spaces into advertising and informa-tion surfaces. The degree of freedom of thought and movement will determine whether, and if so to what extent, a particular and individual appropriation of public space is possible (see Fig. 15).

Fig.16:
A public space surrounded by uniform,
stereotypical, and monofunctional
buildings

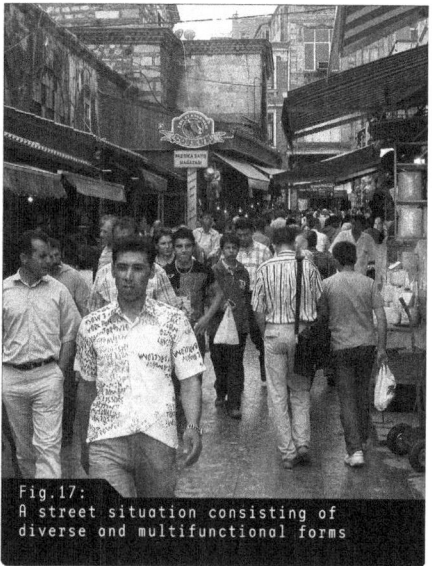

Fig.17:
A street situation consisting of
diverse and multifunctional forms

A private room and an apartment are typical private spaces. Their design is largely based on human scale and is defined by activities and objects that are either partially or not at all shared with the general public. This spatial type frequently possesses a solid spatial shell that clearly separates the interior from the exterior and provides a retreat, safety and security, familiarity, and intimacy. It is constructed to include closable openings, which allows the resident to control who may enter and to what degree.

📎

\\ Note:
The ever-increasing demand for order in the urban configuration results in uniform struc- tures that are frequently assigned only one function. These spaces are being appropriated less and less by different users (see Figs. 16 and 17).

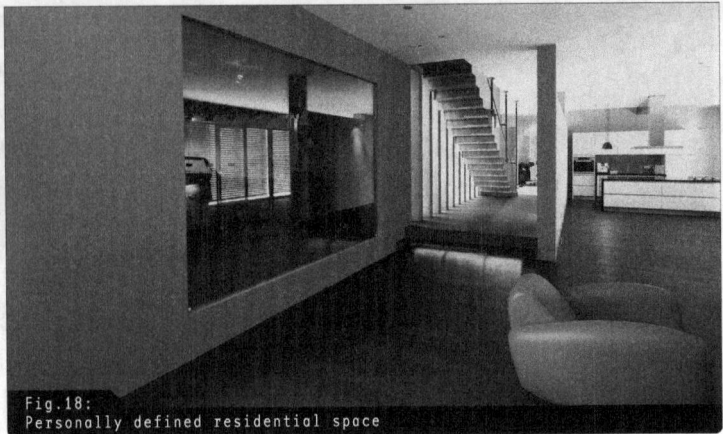
Fig.18:
Personally defined residential space

RESIDENTIAL AND WORK SPACES

A particular characteristic of residential space is that it can be designed primarily according to personal requirements—yet a distinction should be made between shared and individual needs. Needs that are shared by the majority of people are primary needs, which include security, a roof over one's head, and a place to wash. Individual needs go beyond these general needs and are aimed at a type of self-discovery and frequently a self-expression within one's own four walls. > Fig. 18

Individuality and intimacy

Residential space reflects the personality of the resident; after clothing, it is the spatial shell closest to the body. For this reason many elements and materials in the residential space are chosen because they are pleasant to touch. Like a nest, it offers intimacy, warmth, and protection. The residential space is subdivided into zones according to function. Visual associations, spatial divisions, and openings are some means of dividing the space into closed and open areas. The rooms allocated for personal physical needs, such as the bedroom or bathroom, are primarily closed to visitors or the public. They have only a few openings that are small, or difficult to see from the outside, and are usually placed further away from the entrance than the living rooms. Other rooms, however, should be open to both friends and strangers and can serve as a way of expressing or presenting one's own personality. If living and working take place under one roof, it is possible to make the public space a part of the living space, or conversely the living space a part of the public space.

Special residential spaces

In addition to private spaces, there are places such as hospitals, care facilities, retirement homes, orphanages, and hotels that fulfill the

24

special needs of specific social groups. Illness, for example, needs to be controlled and isolated by spatial borders in order to safeguard the physical wellbeing of the majority. On the other hand, these facilities also spatially isolate physical decline and death. Hotels offer temporary shelter to travelers and can be used as residences or spaces for private or professional events.

Work spaces

> ρ

Workshops, production halls, and offices are spaces that are designed according to specific work processes and procedures, the products to be manufactured, and the machinery required.

There needs to be sufficient light, air, and room for movement for the workers or employees, especially because this can help maintain a good level of work performance. There are different spatial types according to the number of employees.

Production lines require spatial shells that are large enough to accommodate their needs. They are designed according to function and the size of the machines, and are not built to encourage people to spend time in them. On the other hand, workshops in the craft trades usually refer functionally to the activities and needs of their skilled employees, because workers here play a greater role in the production of goods. Work places that are publicly accessible, such as a department store, are primarily based on the needs and requirements of the customer.

Office work places are designed primarily for intellectual activities, and the need for movement is minimal. Since the spatial requirements of this work are fairly similar, the typology of office spaces does not vary greatly.

The aim and function of a commercial enterprise is to make money, which means that the design of workplaces is a matter of efficiency in cost and function. To avoid adverse effects on the wellbeing of employees, many countries now have laws to regulate the design of work and social spaces, taking into consideration occupational health aspects that are essential to health and safety. Since offices and factories are designed for a changing staff, there are not many possibilities for the users to give their office or workspace a personal touch.

ρ

\\ Example:
The supporting structures for printing works have to be able to support the weight of the machinery. They need to be strong enough to prevent the building from vibrating while printing is in progress.

CULTURAL AND LEISURE SPACES

Cultural and leisure spaces are for games, spectacles, ceremonies, shopping, exhibitions, and other events that are not a part of daily life and work. They are designed for a large number of users, are usually tall structures that indicate their function and which can be seen clearly. Spatially, they reflect the desire of the public to temporarily abandon the familiar everyday world. They are publicly accessible, provide many spatial attributes, and allow social activities to take place that would normally not be possible in a private or work space. A city park, swimming pool, or landscape outside of the city are typical leisure spaces.
> Fig. 19

Religious ceremonies and visiting churches, mosques, or temples provide spiritual inspiration, which is enhanced by the appropriate spatial type. Museums, theaters, and libraries are places of education, yet also fulfill a communicative and social function. Even commercial adventure parks or shopping centers, as semi-public spaces (restricted opening hours, only a paying public), provide a diversion from the everyday. However, they are primarily dominated by private business interests.

Contemplative spaces

Some leisure and cultural spaces are meant for contemplation. They are usually designed for concentrated, long-term use, for example schools or universities. The contemplative space thrives on special proportions, materials, light, and color. Its unique ambiance can be experienced by everyone. > Fig. 20

Sacred spaces

Sacred spaces are also contemplative spaces with a sublime atmosphere that virtually everyone finds mesmerizing. It is immediately perceived as pleasant and people's behavior adapts to the spatial impression, for example by lowering their voices. Emotional reactions should be deliberately evoked in sacred spaces, in order to make it easier to convey and to concentrate on the religious substance.

\\ Example:
Reading rooms are spaces for quietude and meditation. Here, too, everything is subject to one single function: concentrated reading and study. These spaces provide private areas for temporary use, yet they are also public spaces that exercise strict control over social activities.

\\ Note:
A sacred space's rituals, its special acoustics, unusual spatial dimensions, and often its smells can trigger memories in the visitor of similar spaces he or she has previously experienced. These parameters can be adapted to serve different functions and purposes.

Fig.19:
Different leisure spaces and different approaches to their design

Fig.20:
A contemplative space for concentrated reading

Fig.21:
Archaic structure and interior space inspire contemplation

Fig.22:
Annual religious pilgrimage to the Kaaba in Mecca

Using spatial effect to influence emotions is also a strategy applied when designing spaces that do not have a religious function, such as government facilities, prestigious reception halls, or corporate seminar rooms.

MOVEMENT AND CONNECTIONS

Many spaces are defined by horizontal and vertical travel and traffic routes. Hallways, corridors, stairwells, streets, underpasses, tunnels, and bridges form the traffic zone spatial type. There are different forms of traffic zones for people and their means of transportation, according to the type of use, destination, and speed. In most cases, their directional course is clear. Their function as traffic zones defines a large percentage of the spaces between the architectural structures in a town or city. Stairwells, ramps, or elevators handle vertical movement.

Some traffic zones are entirely dedicated to establishing efficient connections between places, and others also provide certain amenity values. Public streets commonly provide both. The wider they are, the more pleasant they are to spend time in. Squares and intersections are non-directional traffic zones because they have the capacity for diverse

Fig. 23:
Traffic zones / junctions

routes and pathways and can also accommodate other public functions. Functionally defined traffic zones often create resting areas for additional, temporary uses.> Fig. 23

Mobile spaces   In addition to containers, which can be placed on different sites for temporary living or working, cars, airplanes, cable cars, ships, and trains all fall into the category of mobile spaces. Their design is not primarily site-specific and it centered on function, the type of forward motion, and safety requirements. The period spent in these spaces is limited to the time it takes to travel between two places. The interior of a car, for example, can also be a place where longer periods of time are spent, such as on a long journey or in a traffic jam. For this reason they are designed with soft upholstery, textiles, leather, and entertainment electronics to give them a homey and comfortable atmosphere. At the same time, and more than with fixed spaces, the need for status is very important because these mobile spaces are seen more often than fixed spaces by others and at different locations, or are used by several people at a time. There are several fixed-location spatial types for mobile spaces, including stations for parking, service facilities, and pickup or drop-off zones at train stations, gas stations, tram stops, parking

29

garages, bus stations, airports, and so on. Their design is centered on the particular means of transportation, but also take into consideration the departure and arrival aspects of these spaces.

PRESTIGE

Both public buildings and apartments represent attitudes that the owner or resident displays and uses to communicate with, repel, or otherwise influence the user or visitor. Theaters, churches, town halls, and political party headquarters are spaces that reflect their symbolic substance largely by means of architectural or interior architectural means. Select materials in town halls are meant to represent a community's sense of dignity; political party headquarters with glass facades metaphorically communicate to citizens the essence of transparence; courthouses symbolize, in addition to their functional ground plan arrangement, the state's claim to power; and theaters contribute atmospherically to the imaginary world of the stage. ⟩ Fig. 24

Often, power is symbolized by the effect of a large building, or by having small, neighboring structures at a distance.

In everyday architecture, indications of standing or prestige are not as obvious as in public buildings. Many spaces are created that follow ideas conceived by planners for a certain target group and are not

Fig.25:
Sumela Monastery, a structure of permanent, centuries-long use; and a tent as an example of temporary use

directly tailored to the needs of the individual user. Moreover, economic factors or budgets do not often allow housing environments to be personalized, which often results in homogenous residential types. The user is an abstract dimension in the mind of the planner. The lifestyles of the various residents are secondary to the planners' decisions. The planner represents and imposes his or her vision onto a third party. By contrast, residents try to fulfill their needs for self-representation in their homes by means of individualized spatial appropriation. Selected furniture, interior design that is as individual as possible, such as special curtains, or a particularly distinctive entrance doorway, are all ways of expressing individuality. › Chapter Elements and means of spatial design

Spaces represent various philosophies of order, violence, control, and power. Prisons, closed psychiatric facilities, and sometimes entire countries are some places that limit or control the freedom of movement of their occupants.

PERMANENT AND TEMPORARY USE

Spaces can be classified according to their duration of use, because they influence a spatial type as early as the construction phase. A sturdy form, durable material, and a solid construction are used to design permanent structures such as monuments, bunkers, and mausoleums. Wear and tear of the material or various subsequent design additions, which change the original form almost beyond recognition, mean that long use can greatly influence the spatial design. › Fig. 25

In contrast, production halls are constructed only for the presumed duration of the product distribution, and tents are erected for a matter of hours or days. Spaces are decorated or used temporarily for parties, and streets are transformed for the duration of a procession or parade.

31

Fig.26:
Shipping dock assigned a new temporary uses: before and after

Fig.27:
Kokerei Zollverein, Essen: an industrial wasteland used temporarily as an open swimming pool

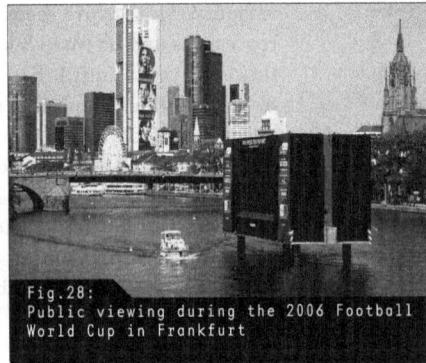

Fig.28:
Public viewing during the 2006 Football World Cup in Frankfurt

\\ Note:
When a large corporation closes down or relocates, it usually has an adverse effect on an entire area of the city. Interim uses for the resulting empty grounds can generate a temporary redesign of the existing spaces (see Fig. 27).

Fig.29:
A ceiling suspension creates a hollow
space of 30 cm

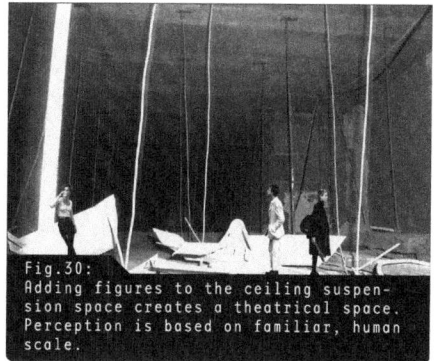

Fig.30:
Adding figures to the ceiling suspension space creates a theatrical space.
Perception is based on familiar, human
scale.

> 🔖

Empty lots in cities or towns are used provisionally and rebuilt to adapt to these new uses. › Fig. 26

Sometimes temporary spatial "implants" for new uses can trigger other architectural developments and processes, and create a snowball effect for their respective context. › Fig. 28 If the new uses prove successful, then the structures that were originally conceived as temporary can become permanent buildings.

STAGED AND IMAGINARY SPACES

Similarly to the spatial types for cultural and leisure functions, stage spaces or set designs also create day-to-day, temporary functional relationships. Imaginary spaces are largely a temporary change of a spatial type. Analogously to dealing with a theater piece, several images are created for the stage and installed and taken down. The audience should be carried away with the help of the constructed scene to an imaginary world for the duration of the performance. The spatial means or lighting effects extend beyond the space and are meant to inspire the imagination. › Figs. 29 and 30

Spatial ideas can be presented and tested using scenographic means. Structures designed for trade fairs or exhibitions are scenographic constructions that allow a company's products or exhibits to be presented or staged in the most appropriate and visually advantageous manner.

Trompe l'oeil

The "trompe l'oeil" effect is a scenographic element used to dissolve the spatial borders of a real spatial type. › Fig. 31

Imaginary
spaces

Architectural ideas outlined on paper, or built three-dimensionally as a model, are imaginary plans for a spatial design that first exists as an idea but has not yet been realized. Many architects use the freedom

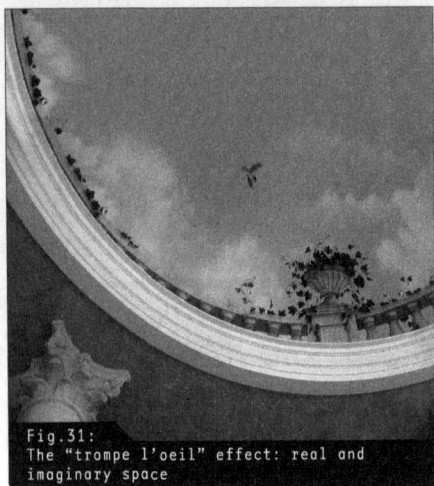

Fig.31:
The "trompe l'oeil" effect: real and
imaginary space

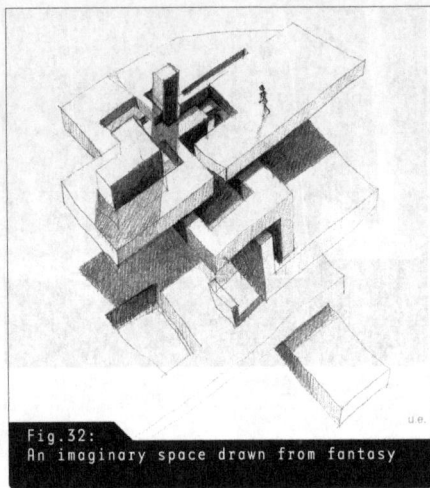

Fig.32:
An imaginary space drawn from fantasy

of imagination to find ideas by temporarily ignoring certain laws such as gravity or outdoor climate. Sometimes this process produces unreal or dysfunctional spaces that seem more like a fantasy than an actual functional structure. › Fig. 32

In addition to function and use, a space possesses other distinctive features that are significant to spatial design and can be emphasized as specific spatial qualities. One fundamental design element in architecture and city planning is the empty space between structures, which can be designed by arranging and positioning individual architectural elements, things, and objects. Space is perceived physically by all the senses and cognitively with our mind; its different dimensions are defined by spatial phenomena. They determine the type, application, and effect of diverse spatial design means that will be introduced at the end of this section.

BUILDINGS IN CONTEXT

Every site has its own specific, spatial environment. The design of a building changes the form of its surrounding space, and conversely, the surrounding space determines a structure's possible design options. A site is influenced by many complex and diverse factors. In addition to the built or natural environment, there are also numerous historical, cultural, and social references that can all serve as contexts. The type, manner, amount, and intensity of the references characterize a spatial design as either contextual or autonomous, if there is no connection or only a weak connection to the features of the environment.

Village, city, and landscape are different architectural contexts, and each determines the type of building design. Neighboring buildings, for example, often set the height of a building's floors, or a ground plan is laid out often according to the available natural light.

SCALE AND SPATIAL DIMENSIONS

Spatial and architectural scale, especially the most standard occurring sizes, is determined primarily by how the structure will be used, and is always perceived in relation to human scale and the adjacent spaces. Proportion is what makes us read a space as large or small. A small building that is based on human scale looks even smaller when placed next to a very large building, because of the great contrast. This also works the other way around. The viewer's individual experience, and the particular spatial dimensions with which one is familiar and uses as a reference, also plays a great role in the perception of scale.
> Fig. 33

A person who grew up in a village with single-story buildings will perceive scale differently to someone who was raised in a city of skyscrapers.

> 🗎

35

Fig.33:
Proportional relationships in a quarry

People are affected by the relative perception of spatial scale effects. It can influence, for example, the way and manner in which we move about a space, or whether or not the space generates a sense of safety and security. A room that exceeds a certain size no longer seems contained, but indifferent; even the effect of scale is lost.

Architects and planners have developed several systems over the course of the architectural history, all which refer to human scale. One of the most recent is Le Corbusier's "Modulor." > Chapter Elements and means of spatial design

INTERIOR AND EXTERIOR

Every spatial border defines a here and now. The impression of interior and exterior is created when additional spatial borders frame a structure in a way that allows a particular spatial depth to be perceived. The spatial shell serves as a communicative channel between the interior and the exterior, and the type and number of openings determine the relationship between the two. > Fig. 34

\\ Note:
Scale is relative, which means that errors of judgment are possible. For example, a piece of furniture seen in a large furniture store might at first appear small and delicate, but later in a smaller apartment, it can suddenly look disproportionately large.

\\ Example:
A large hall with no window to the outside can have a tiring effect on people during a long lecture. A window to the street reduces the degree of spatial closure, and offers a pleasant diversion to the audience and some freedom of movement to the eye and mind.

Fig.34:
A glazed facade as a transition between interior and exterior

The interior of a building can either be clearly visible from the outside or not. A glazed facade creates an almost seamless transition between interior and exterior. The spatial border between interior and exterior can be very thin, as in a glass surface for example, or thick, like the outer walls of medieval fortresses that even had small rooms integrated into them.

Open and
closed

> 🔎

The type and number of openings or the overall permeability of a space's borders determine whether the space is perceived as open or closed. The openings can allow views of neighboring rooms or passageways, both of which also establish an impression of open and closed space.

ORDER AND CHANCE

The given landscape can be viewed as space that has been ordered according to natural influences and conditions, but is often perceived as disordered and chaotic. People organize existing spaces by dividing up sections and delineating areas. Since the topography and vegetation of the landscape are formed according to their own natural laws, any architectural or urban planning project is a mixture or a layering of natural and artificial order.

Chance

In addition to the planned or target-oriented design of rooms, a large part of spatial design will always be left to chance or be subject to the users' organization, due to the many extremely varied and complex relationships between the different elements that shape a room. › Fig. 35 It is therefore recommended to integrate areas where users can influence the use and design into rooms that have been designed according to preset rules of composition.

When building in an urban context, there are usually structures and examples of architecture that were built at various periods throughout

Fig.35:
Building without a planned principle
of order

history and represent earlier principles of order. Constructed order can often still be recognized within the urban fabric centuries later. Old and new order plus chance buildings merge and interweave in many cities, to such an extent that they are difficult to distinguish from one another and create an impression of labyrinth-like disorder.

Imposed or self-determined order

People have to deal with very different systems of spatial order during the course of their daily lives. Many of these have been designed by the city authorities, or by architects, urban planners, and engineers, and not by the users themselves.

One's own apartment is one of the only places that can be at least partially designed according to one's own ideas of order. Private residences convey information about how a person might design his or her private space if free from imposed principles of order. This is where we see the personalized organizing principles of "hoarding." Residents superimpose their own objects and artifacts onto existing spaces and adapt the spaces to suit their private needs. Changes quickly make a difference within this self-determined system of order, so that even the slightest deviation can create a feeling of disorder.

\\ Example:
In some cities new buildings were built on top of
abandoned Roman amphitheaters and adapted over
the course of history to serve other functions,
such residences. The ancient spatial order can
only be seen in the city plan, over which a new
architectural order has been superimposed.

Fig.36:
Spatial depth with the horizon as border

Another, typical spatial principle is the spatial alignments in a room that are necessary for orientation. A vertical distinction is made between up and down for reasons of gravity, and the horizontal axis is defined by the horizon as a constant visible line with a left and right side in our field of vision. All the elements in a room, together with the available lighting conditions, determine the degree to which we visually perceive depth extension. However, movement is what defines depth as a third, spatial dimension, which is what makes space a tangible experience. > Chapter The parameters of spatial design, Time and space

Depth is fundamental to people's perception of space, because physical movement would be impossible without it. The horizon is the constant horizontal spatial border in depth. It is perceived as infinite because it can never be reached or touched. > Fig. 36

Aligned spaces motivate visitors to move along their main directional axis in order to experience the room's dimensions. The alignment is perceived by the senses and the cognitive system in relation to the body, which can move the farthest along this axis. Spaces that are not aligned, for example an inner courtyard or a city square, do not impose a particular direction on the visitor. Rather, they invite him or her to linger, as long as they are large enough and sufficiently lit. Spatial orientation requires a space's borders to be effective; therefore, the more openings a space has, the less effective the directional effect.

DENSITY—EMPTINESS

A space can be filled with things in such a way as to make it seem welcoming and open, or claustrophobic. The feeling of not being able to move within a space is frightening. And conversely, an empty large room where one can move freely can also seem threatening and empty. In such

Fig.37:
Urban density is created by diverse, rich
design and a myriad of human activities.

a situation it is difficult to position one's own body, because things that
the eye would otherwise use to measure and assess distance are missing.
Hence, the size of the room is not clear in relation to the body or objects,
which is the standard criteria. There are few ways in which the senses,
the body, or the head can make a connection or enter a relationship. A
certain amount of spatial density is necessary for people to feel a sense
of wellbeing.

\\Example:
The first guests to arrive at a private party
often linger in the kitchen: a small room, full
of different objects, in which a variety of
activities take place, including entertaining
guests. This room is often more appealing than
the spacious living room, the room originally
intended for receiving guests.

\\Note:
Signs of use such as scratches, chips in cor-
ners, or yellow wallpaper in a room are an
indication of the duration of use and the residents. They give a feeling of authenticity to
a space.

Fig.38:
The signs of time caused by age and deterioration of material

Perceiving a space as being full or empty is always individual and in relation to one's own body size, experience, mood, and freedom of movement. The feeling of fullness or emptiness can also be triggered if a space has many memories for the viewer. The degree of spatial density is very quickly and directly conveyed as either pleasant or unpleasant, and cannot be measured. It is evaluated by personal experience, cultural influences, and physical and mental freedom of movement. > Fig. 37

TIME AND SPACE

Space is always experienced in connection with time. Moreover, visiting a different place always occurs at a different point in time. And at the next moment even this place has changed because of a change in lighting, perhaps the visitor's attention has shifted, or things in a room have been moved to a different position. Walking through a spatial structure allows one to experience time and space, because the spatial sequence might be conditional upon speed, or might have to be completed within a certain amount of time. Since time and space are the defining factors of human existence, our memories are often supported by remembering a certain space or room, and vice versa.

The physical-material nature of rooms is defined by time, because all material ages and its consistency changes over the course of time, which can be a result of sunlight, mechanical abrasion, or simple wear and tear. > Fig. 38

Spaces are witness to past eras and often consist of many elements that originated at very different periods in time. Hence, the condition of the space is a visible sign of the passing of time. > Chapter Types of spaces, Permanent and temporary use

SPATIAL CONDITIONS

The particular spatial effect is determined by several physical and chemical conditions, including temperature, humidity, room acoustics, light, and smell. All of these conditions are typical spatial attributes that work together, change with time, and, most importantly, are all perceived by the close senses. The effect the spatial conditions have on the visitor is determined by the spatial shell's qualities. This is a more or less permeable membrane between the interior and the exterior, through which, for example, differences in temperature are regulated. In turn, the human skin also functions as a membrane between the body and its environment, and is able to sense even the slightest change in temperature or humidity.

Room temperature has a direct effect on users and is both planned and perceived according to human body temperature and activities. For example, office work would be extremely impaired in temperatures under 18° C, yet physical work at this temperature is much more pleasant. High temperatures even make certain forms of physical work impossible. Even clothing, as an additional skin, influences the effect spatial
> 𝕚 conditions have on the body.

The humidity of a room is directly related to the room's temperature; warm air can absorb more humidity than cold air. After reaching a certain temperature called dew point, water vapor condenses into liquid that settles on surfaces in the room with temperatures lower than dew point.

The surface of the spatial borders also significantly influences a space's acoustics. They reflect sound waves or absorb them, depending on their particular surface properties. Sound-absorbing walls do not reflect the waves; the sound waves penetrate the material and are ab-
> 𝕚 sorbed.

𝕚
\\ Note:
A surface creates a sense of discomfort when it is clearly colder than the air temperature and the human body. It draws heat from the body, leaving a sensation of cold, because the temperatures of the room's surfaces are in constant exchange, and compensate for their differences in temperature.

𝕚
\\ Note:
Hard, impermeable surfaces will strongly reflect sound waves, making it difficult or even impossible to hear correctly when inside the space, because of the long reverberation time. The waves are reflected several times and hardly absorbed. If the volume is too high, it can exceed the human ear's pain threshold, making it impossible to remain in the room unless soundproofing measures are taken.

Fig.39:
A spatial opening determined by material

Natural or artificial light is an essential element of spatial design and a fundamental source of information about the dimensions and quality of the space. If there is not enough light reflecting from the space's surfaces, its borders, and hence the space itself, are unclear; and conversely, light can be perceived only when it has a surface from which to reflect. The spatial borders' surfaces reflect incoming light at various levels of intensity and provide information on the dimensions of the space. Spatial depth is enhanced by colors from the blue end of the light spectrum, and by keeping contrasts to a minimum.

MATERIAL

Material dictates spatial construction options and therefore influences spatial form. Materials also determine how one approaches certain details of workmanship, span width lengths, and the type and quality of the space's borders. Therefore the dimensions of the openings in a space and the length of a column-free ceiling are conditional upon whether the structure is a solid or a frame construction of timber, steel, or concrete. › Fig. 39

As mentioned above, a particular material's texture and composition, or color and smell affect the impression and appearance of a space. At the same time, the available dimensions of the structural elements, their construction and hence the spatial form itself are always conditional upon the chosen material and the particular options the material allows for workmanship and production.

ATMOSPHERE

Comfort, coziness, and wellbeing are spatial effects that cannot be truly gauged, but are perceived immediately. Atmosphere is a typical and tangible spatial phenomenon. A room's atmosphere addresses the entire range of human senses in a very direct and complex manner, and eludes rational comprehension. The effect often referred to as "wellbeing" is very difficult to define, partly because sensing it is so subjective. In addition to functional and intellectual needs, atmosphere exists as the focal point wherever people are found. It is defined by human activity as well as by all of the spatial parameters and qualities that the senses and the mind can perceive.

ELEMENTS AND MEANS OF SPATIAL DESIGN

The section below introduces the means and elements that are available for spatial design, the manner in which they may be implemented, and their respective effect. As mentioned in the previous section, atmosphere is created by fulfilling the functional, aesthetic, or technical requirements. It is the sum of diverse spatial phenomena and activities along with their lasting and complex interrelationship, which we perceive cognitively and sensorially. An atmosphere can be inviting or repelling; it can inspire certain activities, trigger memories, or make a visitor want to linger. Its basic nature is essential to our sense of well-being, the type of spatial design, and how we behave in the space.

IDEAS AND CONCEPTS

The idea for a design begins with employing all of the means and elements that shape a space. It can be developed experimentally, determined by the results of analyzing the given parameters, or discovered intuitively. In general however, all three approaches collaborate when creating an idea. › Fig. 41

The design idea is a spatial concept that outlines an arrangement of the users' spatial requirements and how they can be realized architecturally. It serves as the basis for implementing design and its elements.

Concept
Use-related and aesthetic ideas influence the architectural or urban concept. It is the chosen principle of order, which is fundamental to spatial design, and can be characterized by an unusual plan, a special supporting structure, axial references to the environment, a typical sequence of users' of movements, or a special arrangement of spaces and uses. › Fig. 42

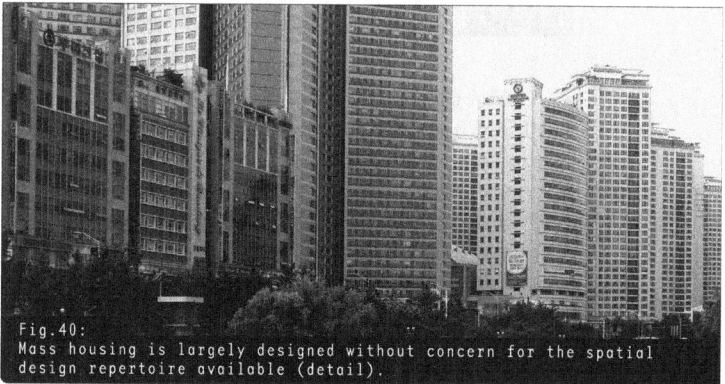

Fig.40:
Mass housing is largely designed without concern for the spatial design repertoire available (detail).

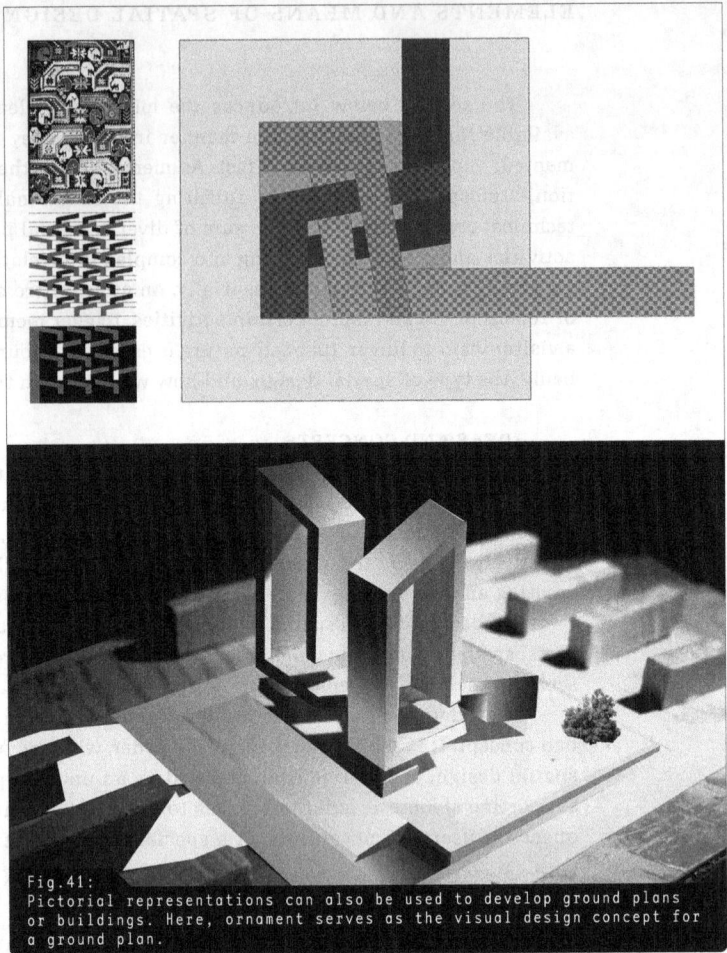

Fig.41:
Pictorial representations can also be used to develop ground plans
or buildings. Here, ornament serves as the visual design concept for
a ground plan.

After the idea has been conceived at first often intuitively or experimentally, drawings and models are developed to test whether all of the factors have been satisfied and issues suitably resolved. This can lead to a series of further attempts until the correct solution has been found.

Since many of the means and elements that shape a space are variable, such as color, light, sound, texture, it is wise to have an overarching spatial concept that forms the design and allows the spatial idea to be clearly recognizable. The concept is a lasting means of spatial design that is typical for a spatial structure and will still be recognizable even if the space is assigned a different use at a later point in time. > Fig. 43

Fig.42:
An existing house is rebuilt using a second house: the concept of a house within a house

Fig.43:
Spatial concept: a city square is characterized by a three-dimensional structure consisting of lines, flat planes, and volumes. Its spatial volume also supports the building, which consists of small sections, and enhances its formal concept.

\\Note:
Additional information and inspiration for developing concepts can be found in: Bert Bielefeld and Sebastian El khouli, *Basics Design Idea*; and Kari Jormakka: *Basics Design Methods*, both Birkhäuser Verlag, Basel 2007 and 2008.

Fig.44:
The temporary use changes an existing environment. Five years ago a small market was built here as a lightweight construction in an empty lot.

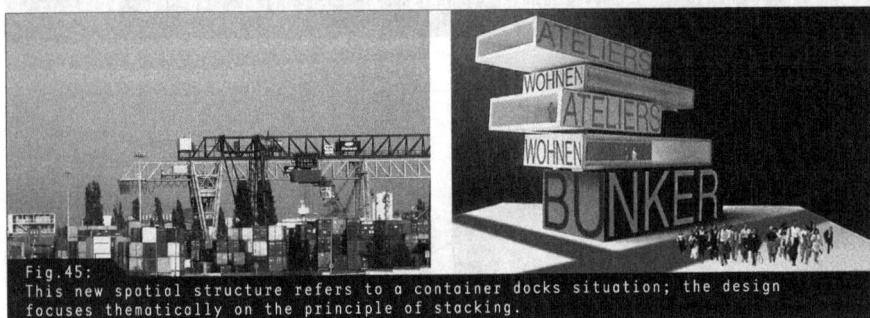

Fig.45:
This new spatial structure refers to a container docks situation; the design focuses thematically on the principle of stacking.

Use

As mentioned above, different uses such as residential, office work, or industrial production shape the spatial type and concept. In the case of existing structures, a specific spatial use is usually prescribed, but it may be newly defined or invented and can as such be used as a means of spatial design. If a new and unusual spatial program is realized in a space that was originally conceived for one specific use, the user group and their activities will also change the spatial design. > Fig. 44

The spatial effect is also altered when an existing site is inhabited by new or different people who are not clothed the same way as the previous users, are present at different times, or move differently. The things they bring with them also affect the space, and temporarily redesign and adapt the environment to their particular needs. A new or different use can also serve to raise critical questions concerning traditional or no longer contemporary functional references.

Context

The environment or an existing building's given attributes form the context. An architectural structure can use these site-specific atmospheric qualities or features as the design's points of reference. > Fig. 45

48

Urban, historical, or social situations can serve as contextual references. The existing uses and situations of a particular environment, such as shopping opportunities or a busy street as a source of disturbing noise, all influence planning a spatial program, and thus, the design of a new construction.

Whether and/or how a building can produce individual contexts from spatial relationships is also a means of design. The context can be obscured by a closed, solid wall, or strategically included by making openings in the structure. In this way, a window with a view to the sea creates a reference to a landscape. If this is missing, part of the environment is obscured, thereby denying a reference to this aspect of the context.

The contextual reference can be read in the form of a building, in its material qualities, or in its spatial program. Using local building material is also a reference to the environment. The color and texture of a building constructed in this material resembles neighboring structures and is well assimilated. In addition, the dimensions or building form determine whether the structure will be well assimilated by or form a contrast to the given context.

SPATIAL NOTATION

Every spatial design first requires knowledge of the site or existing spatial situation that is to be designed. The attributes of the terrain or existing building are investigated and recorded when inspecting the site. The term used for this procedure is spatial notation. It comprises the quantitative and qualitative data taken from every spatial quality, which are sketched, noted down or recorded in written form, photographed, documented on film, and measured. The spatial extension and attributes of the given site are calculated and documented as objectively as possible by means of measuring equipment. These data are then used to develop a strategic approach using the available spatial design means. In addition to traditional measuring tools such as tape measurers, folding rules, and barometric levels, digital laser measuring equipment allows a precise reading of the spatial extension in all three dimensions.

\\ Tip:
Because a folding rule or other appropriate measuring devices might not always be available, it is good to know the length of your stride or the width of your hand in order to roughly calculate distances without the need for measuring equipment.

\\ Note:
Precise calculations and measurements are usually modified to suit the design task. Building furniture requires a different dimensional accuracy than designing a street junction. These dimensions are accepted as objective and veritable for the planning, but are often approximations.

Tab.3:
Common formats used in presenting different spatial scales: 1 cm in plan =
x cm in reality.

Landscape	City	Building	Blueprint for executing construction work	Furniture	Construction detail
1:100,000– 1:2000	1:10,000– 1:500	1:500– 1:100	1:50–1:20	1:20–1:1	1:10–1:1

Spatial notations and documented technical data convey the site's spatial attributes as well as the different spatial design concepts to users, workers, and all those involved in the construction. Since the space can usually only be viewed individually and directly on site, this method allows the designer to present the space and thus convey information about its qualities off-site, so that others will be able to structurally realize the spatial design idea.

Full-scale
illustration

As a work basis for the spatial design, spaces are presented in a scaled-down form. The scale is chosen so that the presentation format (for example a printed piece of paper) is easy to handle but also large enough to show and process the necessary detailed information. › Tab. 3

All lines in the drawing are abstractions of the real space's attributes because they only represent actual surfaces and materials that overlap or are adjacent to one another. Spaces are usually visualized in two-dimensional drawings such as ground plans, sectional views, and elevations. These orthogonal projections of spatial bordering surfaces are the fundamental notation elements for visualizing, defining, and communicating space. › Fig. 46

Orthogonal visualizations provide details about the geometric and dimensional details of the space. At the same time, they abstract and simplify its complexity, because they can never provide the complete information. Other analyses and notations of spatial attributes such as sound absorption, the material qualities of the spatial surfaces (texture, color, material), site inspections at different times of day, or the knowledge of the site's history are all useful for realizing a spatial design concept. For spatial aspects such as these, there are usually only a few appropriate means of notation available, such as instruments that measure light intensity and noise level. Discussions with users and neighbors can also often provide valuable and unexpected information about the site-specific conditions.

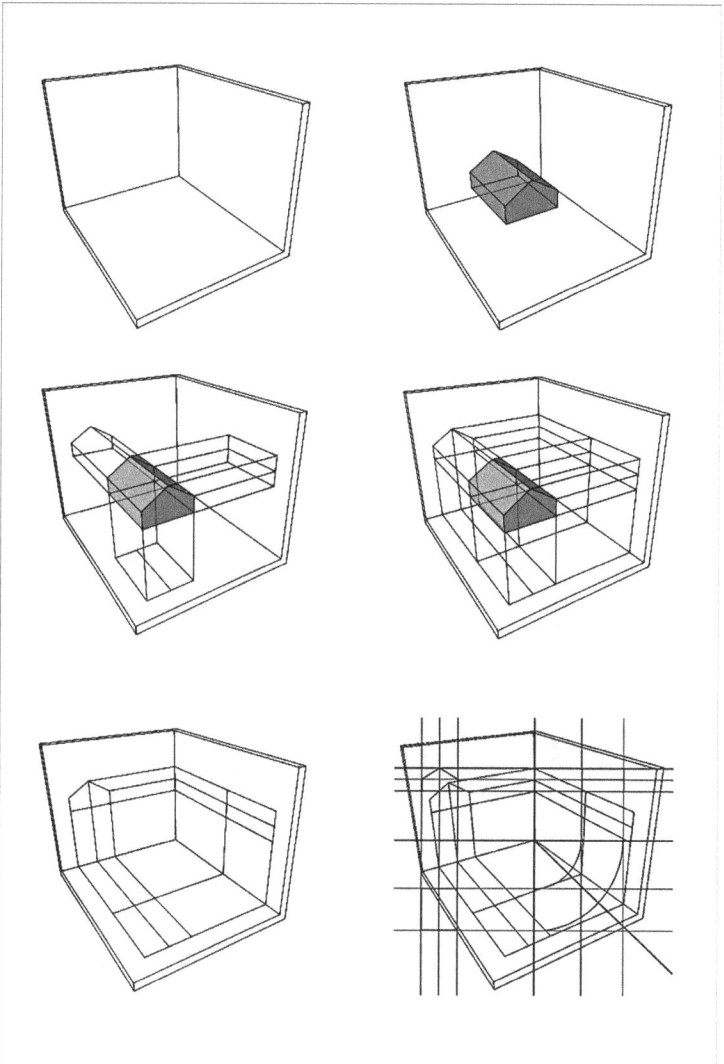

Spatial
situation

In order to check or convey a spatial design idea, the space is presented on film as a set visual image, either in perspective or three-dimensionally as a model. A layperson may often not understand orthogonal projections. They are more familiar with perspective images

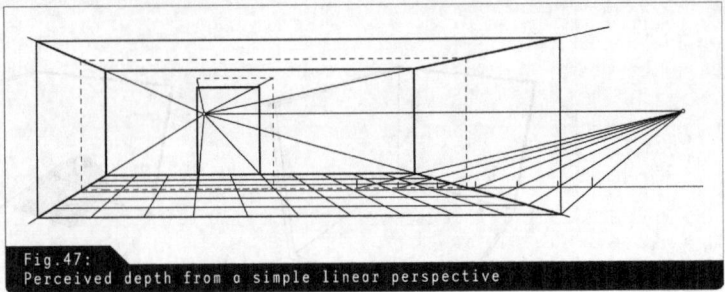

Fig.47:
Perceived depth from a simple linear perspective

Fig.48:
Spatial presentation using projections of structural images onto the model to check and assess the spatial intervention.

and the scaled-down model as forms of spatial visualizations, which are therefore more helpful in conveying spatial design ideas.

A perspective drawing can illustrate a real three-dimensional space precisely, photo-realistically, or abstractly using only a linear framework. The three-dimensional space is projected onto a screen; all of the depth contours run diagonally to one or more points in the horizon. › Fig. 47

Since viewers have an individual ability to abstract, a too-precise photo-realistic presentation might give clients or users the impression that the design idea is already complete and that they can no longer contribute to or influence the concept. A hand drawing can also convey an idea well and provisionally enough so that changes can still be made.

Space can now be simulated using powerful computers and given the impression of movement within a three-dimensional space. The viewer's position and eye level can be adjusted, thereby reproducing his or her everyday experience of space.

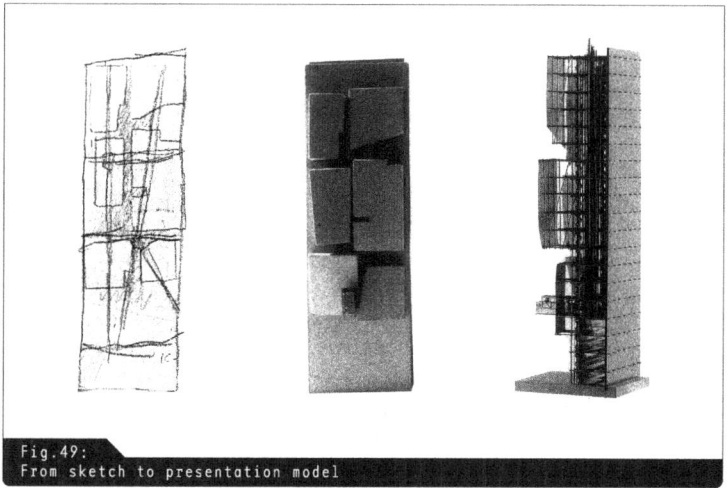

Fig.49:
From sketch to presentation model

Spatial models

In addition to perspective visualizations, models are also very common tools that can simulate and present space on a smaller scale. Models are also very helpful in understanding spatial relationships, dimensions, as well as in planning and conveying spatial ideas. Models allow a better comprehension of space and a direct communication of ideas, because they are the most similar to day-to-day spatial experience. The viewer's position and eye level move freely in relationship to the model. In addition to the visual aspect, models can also communicate the haptic impression of individual materials or make it possible to test various light qualities. › Fig. 48

Models of spaces or spatial components, such as a facade element, are built on a scale of 1:1 in order to test and discuss a planned construction. › Fig. 49

› 🗋

🗋
\\ Note:
Further information and inspiration regarding spatial visualizations and communicating spatial concepts can be found in: Bert Bielefeld and Isabella Skiba, *Basics Technical Drawing*; Jan Krebs, *Basics CAD*; Alexander Schilling, *Basics Modelbuilding*; and Michael Heinrich, *Basics Architectural Photography*, all Birkhäuser Verlag.

COMPOSITION, PROPORTION, DIMENSION

All means and elements of spatial design are put together in a spatial composition. A composition is the product of a designer's strategically compiled and arranged spatial elements. Similarly to the way in which music is composed, structural elements and spaces are planned and arranged so that they relate to one another. Individual spaces or sequences of spaces are composed according to requirements of use, and aesthetic concepts and ideas. The functional necessity of creating a connection between two points is just as influential to a spatial composition as the construction site or the floor space required for a certain machine. The architectural, spatial composition can be based on geometric laws, proportional systems, instinct, a two-dimensional image, axial relationships to points in the environment, or even derived from specific topographic aspects of a landscape.

Order and chance

Most spatial designs and compositions are produced on a daily basis by users' movements and the objects they reposition within a space. Even the scent or voice of a person can greatly affect the spatial impression, and moving elements around in an existing space will constantly change its previous order and composition. Moreover, the specific positions of structural elements can never be completely planned or foreseen, particularly in structures for long-term use. Part of the composition has to remain subject to chance since not every aspect can be controlled 100% by the design. For this reason, many spatial compositions integrate areas that were not designed to be use-specific.

> ℘
Experimental design

Exciting spaces can be created using experimental approaches in addition to rational-analytical methods. Simple geometric bodies can be quickly turned into spatially complex figures by using a few specific procedures. Two-dimensional, rectangular figures or three-dimensional, highly regular geometric forms—such as a cube—can be divided, doubled, folded, or transformed by other geometric laws to become diverse spatial forms. > Figs. 50–52

℘
\\ Example:
A building's future users are mostly unknown to those in the housing construction trade. Open spaces have to be planned for cupboards, beds, chairs, and other furniture. The future design of spaces such as these is far removed from the influence of "spatial composers."

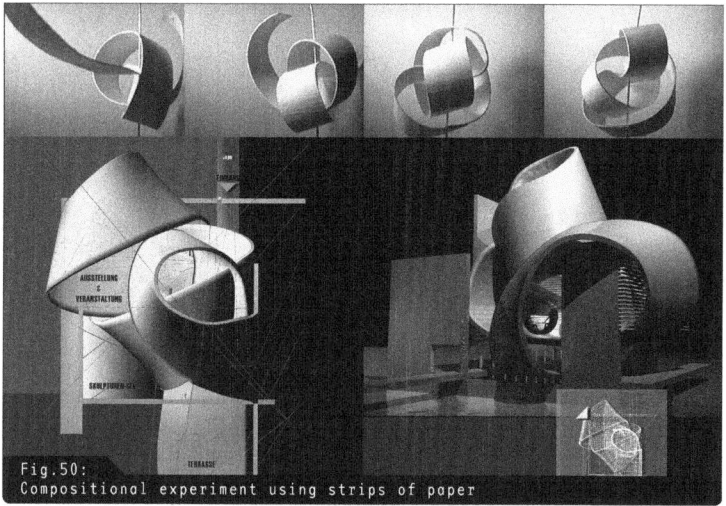

Fig.50:
Compositional experiment using strips of paper

Fig.51:
Compositional experiment using found natural forms

Two-dimensional materials like paper, wood, metal, and so on can be developed to form three-dimensional figures by shaping them into different forms, using methods that suit their particular material attributes (stiff or flexible, for example). These objects, with forms derived from the quality of their material, can be tuned more finely according to qualitative focal points, which include dynamics of tension, play of light

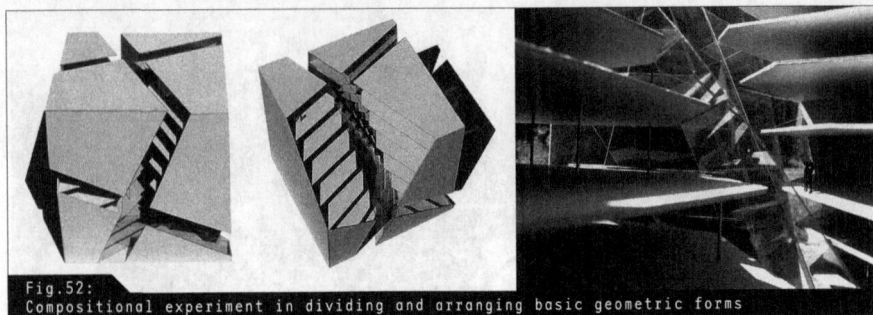

Fig.52:
Compositional experiment in dividing and arranging basic geometric forms

on the surfaces, the randomness resulting from how they were formed spatially, and so on.

Adapting a different manner of presentation and different scale will result, after a further phase of development (for example 3D animation), in a tangible visualization of the spatial object. This process can be taken as far as the pragmatic level of an actual building design. An apparently coincidental spatial configuration becomes a built structure and can still be comprehended throughout this type of process.

Looking at different objects or empty shells produces various personal associations that trigger an idea and set a design process in motion. A model can be a direct means of realizing an idea. Working with a design model makes it possible to immediately control an idea and get feedback from it. The interaction between working with one's hands and conceptual intellectual work builds confidence in assessing one's own work. This exercise can also help develop one's repertoire of realization strategies.

Proportion

Spatial proportions describe the relationship between the width, height, and length of a space or spaces. Certain spatial proportions with fixed geometric laws and dimensions were obligatory in design for centuries. One proportional relationship that was considered well balanced, such as the Golden Section, defined the dimensions of all spatial elements in the ground plan, sectional view, and details. A space based on the Golden Section has a tranquil and balanced effect. > Fig. 53

There have been repeated attempts throughout architectural history to develop and apply standard proportional tenets and systems. Le Corbusier's Modulor was an attempt to regulate the scale all spatial elements to fit the standard proportions of the human body. His proportional tenet was based on the Golden Section and a mathematical series of numbers (Fibonacci Series).

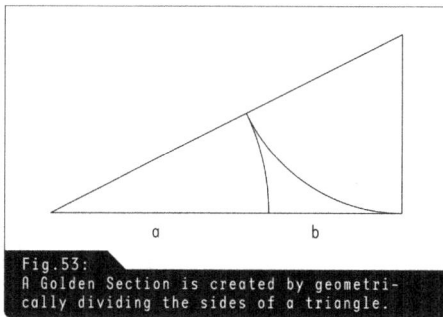

Fig.53:
A Golden Section is created by geometrically dividing the sides of a triangle.

Structural elements can also prescribe proportions. Using bricks as the smallest building component creates a modular grid that can determine spatial dimensions. Spatial proportions in Japan are traditionally calculated and defined by the number of *tatami* mats, which measure 1:2 (85×170 cm). The length is based on the average height of a Japanese person. A traditional Japanese room is the size of 6 mats.

The metric system of measurement, which is most prevalent today, is based on the circumference of the Earth and no longer refers to the human body. However, the "Imperial" measures used in England and the USA (feet, inches, etc.), and some other historical systems of measurement, are still based on the human body.

A space with a square floor area and an equal height will have a very tranquil effect, because all of the space's edges are equal in length. Each individual spatial dimension is always perceived in relation to one another and in relation to the human body. Doubling the room's height

\\ Note:
There are two lengths in relation to one another in the Golden Section, if the ratio between the sum of those quantities and the longer one is the same as the ratio between the longer one and the shorter: a is related to b as a+b to a. This ratio can be seen in nature as well as on the human body. The Golden Section is used in the same manner in architecture, art, and music and considered a balanced and harmonious proportion. In figures the ratio is 1.618:1.

\\ Note:
Some spatial proportions are influenced by industrial mass production and the dimensions of their means of transportation, such as shipping containers or euro-pallets. The dimensions of objects and appliances that have been adapted for optimal utilization, for example the floor area needed by kitchen appliances, in turn influences the design and measurements of kitchens.

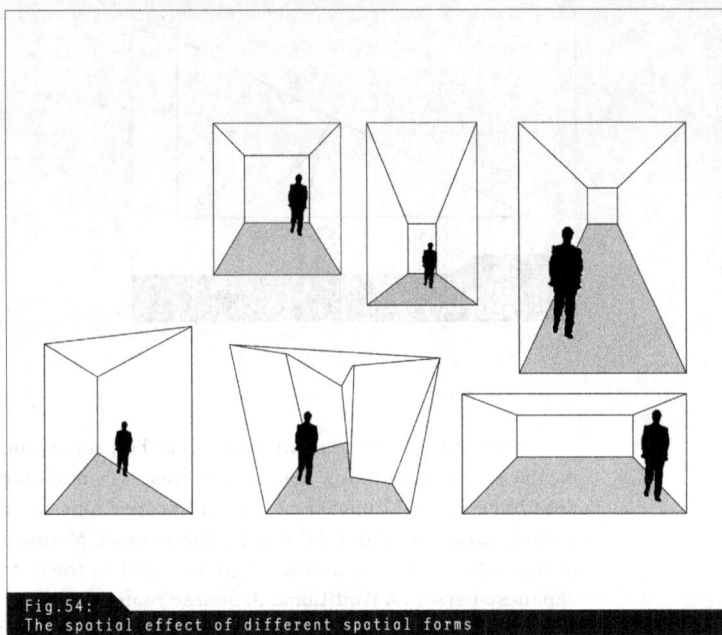

increases the vertical effect. If the floor area is made longer, the room will receive a directional thrust that motivates people to move in one direction. If a ceiling is low, and the room is almost too low to stand in, it is perceived as claustrophobic because it becomes difficult or impossible to shift the body's position. The same is true for a narrow corridor that is wide enough to accommodate only one person. Since this does not provide enough room to linger, people will move quickly through this space looking for an exit. > Fig. 54

Dimension   The massive difference in spatial types and the familiarity with certain common proportions can be used as a design means to create spatial contrasts. The proportions of spaces are always based on one another and on human scale. A building of standard dimensions suddenly looks much smaller than it really is when placed next to a very large structure. And the reverse is also true: a building looks higher when standing alone. If elements that are small, familiar, and based on human scale are integrated into very large rooms, the impressive overall spatial effect is increased. > Fig. 55

Emptiness and   Emptiness and density are essential aspects of spatial composi-
density   tion. Spatial density has little to do with how much a space is filled with

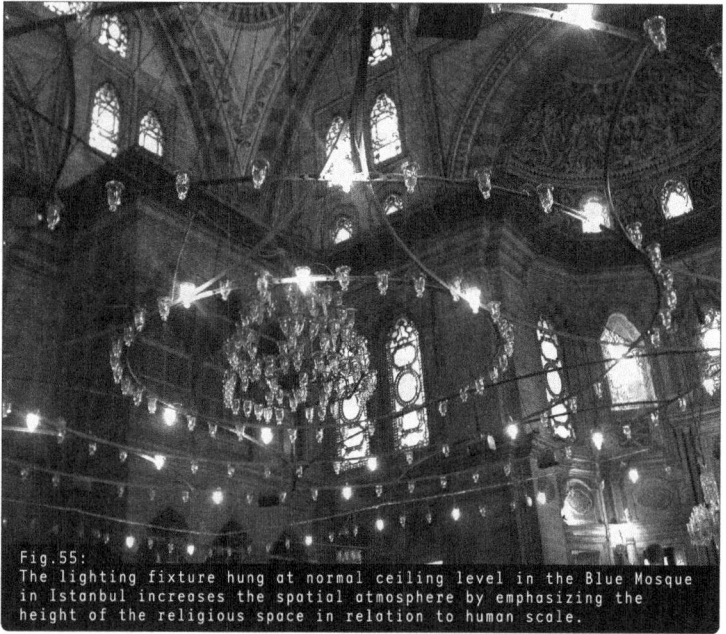

Fig.55:
The lighting fixture hung at normal ceiling level in the Blue Mosque in Istanbul increases the spatial atmosphere by emphasizing the height of the religious space in relation to human scale.

people, objects, associative possibilities, or activities. The experience of the body in space, or in other words the distance between the body and spatial borders or objects, together with our cognitive spatial awareness, determine whether we read space as empty or dense. A person might not be free to choose his or her position in space if a space is experienced as too full, which could result in a spatial impression that invokes fear (a possible reason for mass panic). A certain relationship between spatial density and emptiness determines the individual wellbeing of the viewer, without this being definable or quantifiable. › Fig. 56

Nonetheless, designing the degree of spatial density is used as a means of spatial design, because it has a similar effect on many people. Expanding the distance between objects or buildings will lessen the density of the space and increase the sensation of emptiness. In this case, there are not enough spatial coordinates that, with the aid of measurements and distances, would otherwise help us determine our position in space. Space is created by the interaction of elements and perceivable intermediary spaces. In the desert, in complete darkness, or at sea, spatial borders may be only partially discernible or completely invisible, which makes the space appear empty enough to seem threatening.

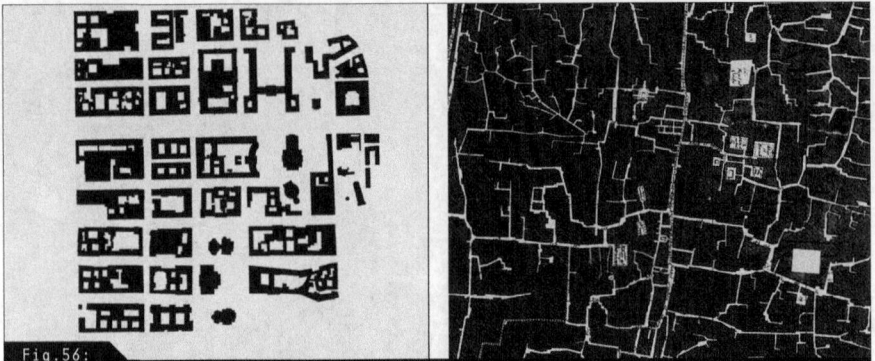

Fig.56:
As-built plans of Berlin and Cairo demonstrate different urban densities with the aid of abstract renderings of the city structures' building masses and empty spaces.

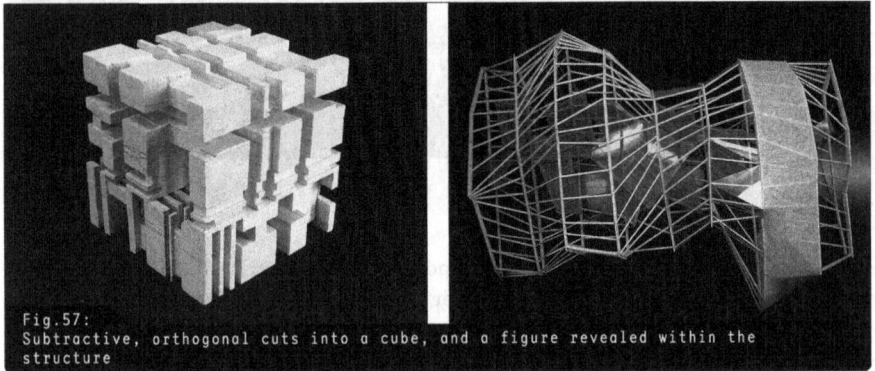

Fig.57:
Subtractive, orthogonal cuts into a cube, and a figure revealed within the structure

SPACE, DESIGN, STRUCTURE

Although various construction or formal aspects are able to determine spatial shape, the tectonic structure that is determined by gravity absolutely does.

Consequently, the support structure of a building can be designed as flat and solid without visible constructive structure, or as delicate and open using supports. > Fig. 57

Spaces with smooth, flowing transitions, or spaces that are clearly separated, as well as integrated empty spaces, smaller cells in a large space, and orthogonal or freely formed spatial borders are all spatial structures that give the built structure its unique form.

SPATIAL BORDERS AND CONNECTIONS

Spatial shell   Creating spatial borders is a basic means of spatial design. Spatial borders divide and zone off the infinite amount of space above the Earth's surface. A spatial shell is created when several spatial borders exists to define a width, depth, and height. An interior and an exterior are created by the two sides of a single linear spatial border. The spatial shell protects against cold and heat, humidity and moisture, noise, and unpleasant or unwanted views. The degree of permeability between interior and exterior is what makes a spatial shell appear open or closed.

This impression of open and closed is determined by the attributes of the spatial borders and the way in which they have been divided, as well as the lighting conditions, and the proportional relationships in the space. Structural elements can provide a three-dimensional quality to floors, walls, and ceilings, and the shadows they create will emphasize the spatial borders.

Spatial connections   Connections between spaces and the interior and exterior are created when openings are made or integrated into the spatial border's surfaces or separate rooms. They provide access or a view to the adjacent spaces and link them horizontally and vertically. The spatial shell is similar to a membrane that is perforated by doors and windows. The number and form of the spatial connections or the permeability of the membrane are important design means and determine the spatial impression. > Figs. 58–62

The effect of the openings in a space relies greatly upon whether the openings offer an exit or only a view outside, or, in other words, whether there is a way to cross the spatial borders and if so, how. Openings in the surfaces of the spatial borders also provide daylight, fresh air, or a change in temperature to the interior. Windows allow a view through the spatial border in both directions. While garages are measured to fit a car, the doors to a space should correspond to human scale, be inviting to guests, and provide security for private spaces by denying access to uninvited visitors. For this reason, spatial openings are often designed so as to slightly vary the degree of openness via doors, curtains, Venetian blinds, and window shutters.

ⓘ
\\ Note:
Gravity and the flow of load forces can either be made visible and emphasized, or made less noticeable than other design features.

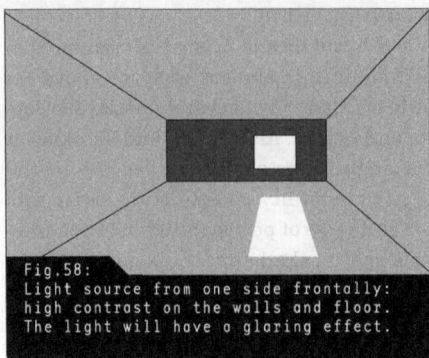

Fig.58:
Light source from one side frontally:
high contrast on the walls and floor.
The light will have a glaring effect.

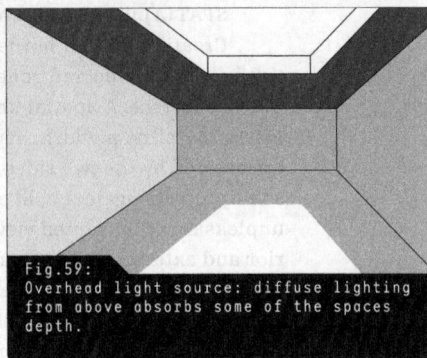

Fig.59:
Overhead light source: diffuse lighting
from above absorbs some of the spaces
depth.

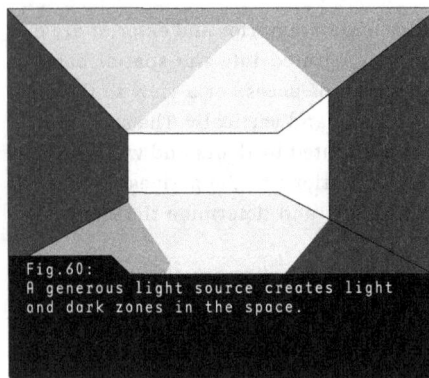

Fig.60:
A generous light source creates light
and dark zones in the space.

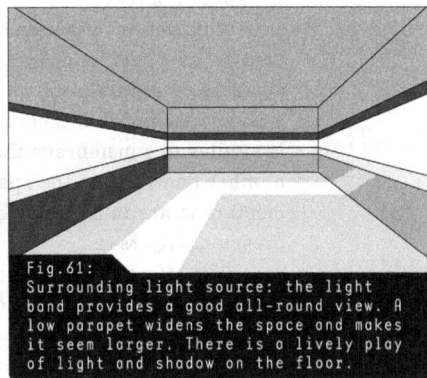

Fig.61:
Surrounding light source: the light
band provides a good all-round view. A
low parapet widens the space and makes
it seem larger. There is a lively play
of light and shadow on the floor.

A space will seem closed and separated from an adjacent space if the wall opening is only 70 × 200 cm in size and has a lintel; in turn, the space's borders will seem to dissolve if the opening is twice as wide and is floor-to-ceiling high—this creates a feeling of two spaces flowing in and out of one another. A seamless transition between the floor surface and ceiling will enhance this effect. A wall opening that extends to the floor will create a more spacious and open impression than a window with a parapet. Ceiling-high ribbon glazing emphasizes the horizontal axis. On the other hand, a ceiling-high glass wall almost completely dissolves the borders between the interior and exterior, which suspends one's impression of being inside and makes the space appear to stretch far beyond its borders. Dissolving one spatial edge to create a view or

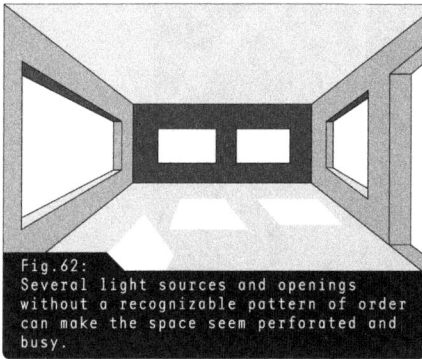

Fig.62:
Several light sources and openings without a recognizable pattern of order can make the space seem perforated and busy.

passage will make the space seem very open and emphasizes the diagonal axis. The dissolved edge undermines the impression of a surrounding shell, makes the construction seem less stable, and creates a feeling of unease.

The position and direction of the openings will structure the walls and ceilings into horizontal or vertical sections. Openings made in the walls and ceilings turn the visible areas of the adjacent spaces into pictures and elements of the space, and the window or doorframes become picture frames.

> 📱
Vertical
connections

Spaces are connected vertically via stairs, ramps, elevators, and ladders, or through openings in floors or ceilings. Vertical connections are the diagonal elements that are necessary in order to move through spaces; they can either stimulate motion or create a sense of unease.

📱

\\ Note:
Door thresholds mark the transition between inside and outside. Throughout architectural history, different cultures have designed thresholds to stress the shift between interior and exterior and make the attributes of the spatial border a tangible experience. A high threshold or a painted threshold emphasizes the spatial zone in the wall thickness between the inside and outside. There are increasing requirements today for building without thresholds—barrier-free architecture—that need to be considered.

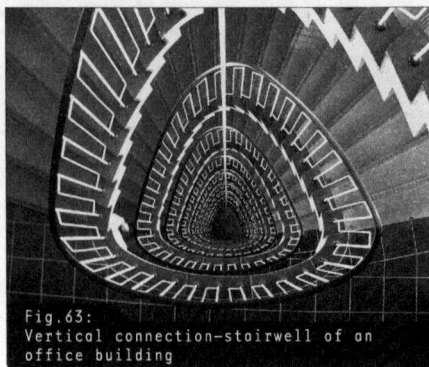

Fig.63:
Vertical connection—stairwell of an office building

The tilted surface of a ramp or the horizontal steps on a stairway are positioned for aesthetic reasons either open in a space or in specifically allocated areas, which helps to control sound transmission or impede the possible spread of fire (staircases, stairways, and elevator shafts). › Fig. 63 The design of stairwells emphasizes and enhances the vertical and diagonal connections used for communication and light.

Very high spaces can accommodate two stories in some areas. If they are open on one side facing the high space (gallery), the entire area will flow together and be restricted to one space with spatial zones of different heights. Another potential connection option would be to restrict the levels by using differences in height, as is done for example with mezzanines placed at mid-floor height.

LAYERING

Wherever the surface of one material or three-dimensional body ends, the surfaces adjacent to, in front of, or behind it are all visible. The type of layering affects a space and determines its depth effect. It can also emphasize changes as well as different points in time.

ρ

\\ Example:
A series of columns divides a long facade into several segments that are closer to human scale than the entire length of the building. The true length of the structure would have an unpleasant effect on the visitor. The columns create a rhythm that invites and guides one's movement through the space along the facade.

Fig.64:
Layering of building components that is either complex or simple and quickly read can enhance a space's effect of depth.

> ᵖ

Vertical
layering

Layering and
transparency

Elements that are arranged sequentially in the space can be designed and positioned so as to influence the spatial depth effect. This occurs with layered building components that are arranged as stacked or layered one after the other, which divide the spatial depth into individual zones or sections and enhances the spatial extension. > Fig. 64

Due to gravity, layering is also an essential spatial design principle regarding the vertical axis. Buildings are made by vertically layering components on top of one another—furthermore, stacking elements accentuates and highlights gravity and the vertical axis. The construction will seem transparent if the interior structure of the stacked or layered elements is visible. Consequently, the facades of high-rise buildings are usually structured vertically using openings, cornices, or parapets, in order to increase the vertical effect.

The history and origins of a space can be read in the spatial layers that have been added at different points in time. It can also be seen on different surfaces and construction elements that vary in age as well. Certain elements of the space provide details about the past and stimulate the cognitive system to make associations and to imagine the age of the space, its historical past, or former inhabitants. The passing of time becomes tangible because these traces of the past are still visible in an existing building after it has been redesigned. Even cities, urban areas, and landscapes are very frequently a result of layer upon layer of various spatial arrangements and chance designs, giving them a more labyrinth-like than transparent impression.

TRANSPARENCY

Perforations in the form of openings, and the degree of their light permeability can reveal what a space's border surfaces are actually meant to conceal. Transparency and concealment are also used as design means

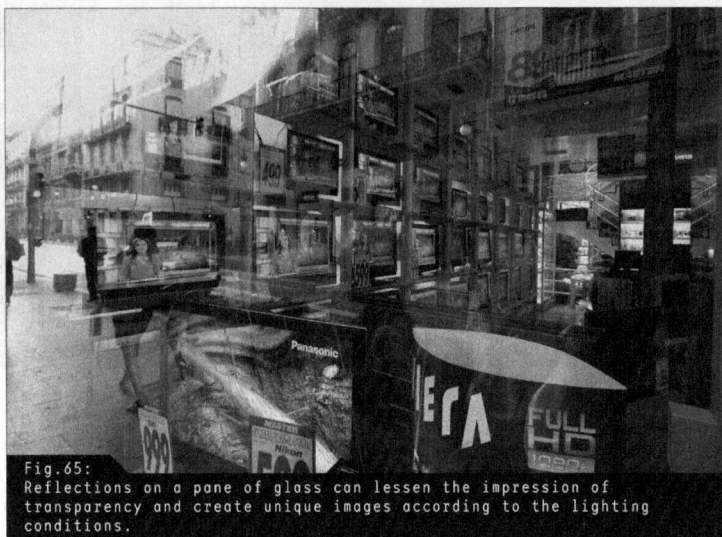

Fig.65:
Reflections on a pane of glass can lessen the impression of
transparency and create unique images according to the lighting
conditions.

in order to control the degree of public access or intimacy and privacy.
While a pane of window glass is transparent (curtains at most reveal contours when lit from behind), a wall conceals everything that is behind it.

The impression of transparency is created when the depth, attributes, and arrangement of spaces or sequence of spaces are clearly perceivable. Spaces also seem transparent if the viewer or user is able to
easily identify their own position in space, to orient him or herself, and
to find the entrance or the exit. The effect of transparency is also created
when a viewer can look inside a building to see what use or function it
conceals, or to see the layout of its ground plan. › Fig. 65

CHOREOGRAPHING SPACE

Spatial choreography is the term used for the spatial design of a
sequence of spaces. It governs the movement and behavior of the user in
the space. Space is generally perceived while the user or viewer is moving through it. Their line of movement is free and but also determined by
the attributes of the space or sequences of spaces. › Fig. 66

One example of spatial choreography that is rich in contrast would
be a round space located at the end of a long and narrow hallway, has no
dominating spatial direction, but a tall stairwell passing through several
stories of the building. This example of spatial choreography might produce an uneasy impression, but it could also be surprising and stimulate
the user to explore the levels above. The antechamber of a church, which

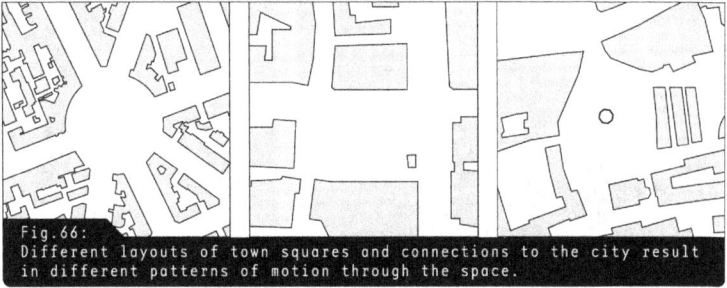

Fig.66:
Different layouts of town squares and connections to the city result in different patterns of motion through the space.

is narrow and has a low ceiling, serves as a transition from the exterior to the interior. It is strategically intended to engulf the visitor and empha-size the spatial effect of the high ceiling in the main hall. How the spatial proportions and attributes of spatial sequences are designed can influ-ence a visitor's behavior, as well as the direction in which he or she moves through the space. Spaces can be staged in such a way as to motivate the user to behave in a manner that corresponds with the designed scene, similar to an actor or actress in a stage play. This can be compared with decorating a space for a celebration in a way that will transport the guests temporarily from an everyday situation into a special atmosphere.

A sequence of spaces can be designed so excitingly that they chal-lenge all of the visitor's senses. The lines of movement through a space on both the level and the vertical axes can be designed as straight, bent, or curved, or so as to encourage one to move quickly or slowly.

Medieval cities frequently seem like labyrinths because they have a rich array of spatial impressions, engage all of the senses at the same time, and challenge our motor abilities. The line of movement here has only a few long linear segments. It is characterized far more by diverse

ρ

\\ Example:
A recognizable sequence of spaces can heighten curiosity and motivate the user to move through them. In a single room with two openings, the user will walk from the door to the window in the direction of the light. A line of movement is also created when there are two door open-ings in an otherwise empty room.

Fig.67:
A varied and a monotonous traffic zone (subway station)

changes in direction, varying spatial proportions, and a sequence of short and long pathways. A medieval square encourages a visitor to linger rather than to move—as long as there are no potentially dangerous cars. Cities today are mostly tailored to a motor vehicle's radius of movement. Urban planning such as this often hampers pedestrians' or cyclists' movement and can be monotonous and tiring. › Fig. 67

Vertical movement is more difficult than horizontal movement because of the force of gravity, which pulls towards the Earth's core. Consequently, comfort, rhythm, and the duration of the specific vertical movement are essential factors when designing stairs or ramps. Platforms placed at intermediate levels provide a place to rest, a place to change direction, and also divide the climb up into manageable segments.

The pattern of movement expected in a space also determines its form. The most important recurring movements are documented in diagrams as form-defining parameters that will serve as the most direct possible basis for the design of the spatial shell—consequently, the direction, amount, intensity, and speed of these parameters define the spatial shape. The building's design is developed as a function of the anticipated movements at different times of day. Sequences of movements are calculated empirically even for furniture, objects, or machine construction in order to establish a form that best suits the movement.

LIGHT AND SHADOW
For people, light is the visible part of electromagnetic radiation. Yet in the context of physics, "light" represents the entire electromagnetic spectrum of waves. Daylight from the sun and artificial light from various electric sources are all reflected off spatial surfaces. Space can only be perceived if the spatial borders and dimensions are sufficiently

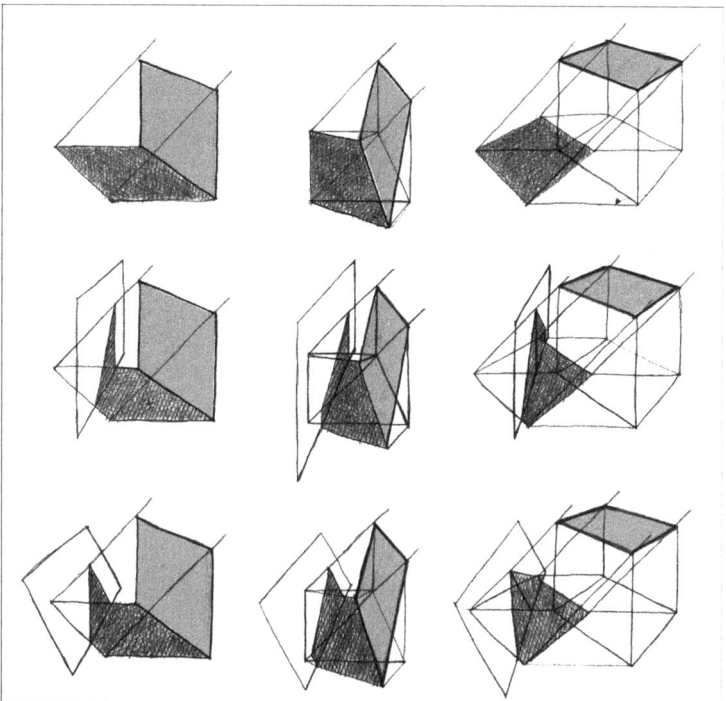

Fig. 68:
Spatial shadows from various surfaces create a strong three-dimensional effect even in the simplest spatial configurations.

visible, hence spatial design is always also lighting design. The surfaces that form a space's borders reflect incoming light with more or less intensity; they reveal the extent of the space and enable a person to locate their point of orientation within the space.

> ⋒

⋒

\\ Note:
The area of light visible to humans has a wave-length of approx. 380-780 nanometers (nm), which corresponds to a frequency spectrum of approx. 789-385 THz. A precise limit cannot be established because the sensitivity of the human eye to the limits of the light spectrum diminishes gradually and not abruptly.

Tab.4:
Wavelengths of the prismatic color of light (in nanometers)

Violet	Blue	Green	Yellow	Orange	Red
380-420	420-490	490-575	575-585	585-650	650-750

Tab.5:
Typical luminous intensity measured in lux (Central Europe)

Bright daylight sun	100,000 lx
Cloudy summer day	20,000 lx
Cloudy winter day	3500 lx
Television studio	1000 lx
Room and office lighting	750 lx
Hallway lighting	100 lx
Street lighting	10 lx
Candles one meter away	1 lx
Full moon at night	0.25 lx

Light would not exist without space, objects, particulate matter, or the humidity in the air, because it would have nothing to reflect off. Light can penetrate textile and opaque surfaces partially or not at all. Shadows are created according to position, intensity, and direction of the light source. > Fig. 68

Light color

> 🔟 > 🔟

Light splits into several spectral colors within the area of the spectrum visible to humans. The associated spectral color is determined by the wavelength of the maximum of the continuous spectrum and given a corresponding color temperature (TCP), measured in Kelvin (K). > Tab. 4

🔟
\\ Note:
The filaments in a light bulb have a relatively constant wavelength and consequently produce a continuous spectrum similar to sunlight.

🔟
\\ Note:
Light is an environmental factor, like sound or exhaust fumes. Light emissions from lighting facilities can harm people and animals or even impair technical processes.

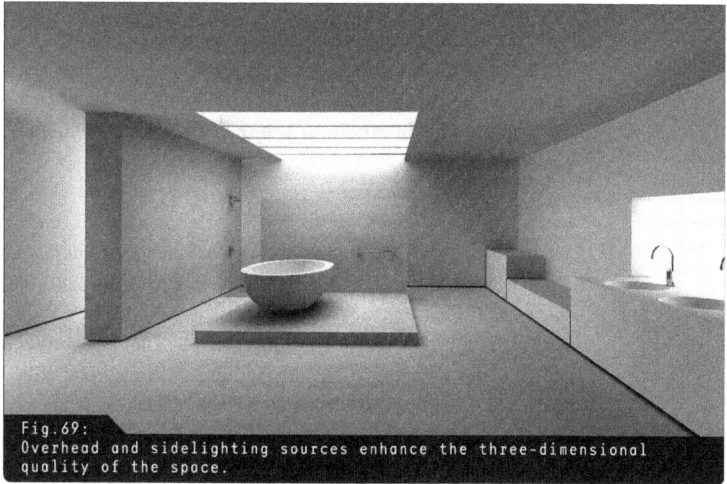

Fig. 69:
Overhead and sidelighting sources enhance the three-dimensional quality of the space.

› Ω

Light is measured in a similar manner to color; perceiving light is very individual and it is difficult to agree. › Tab. 5

Lightness and darkness are essential elements of spatial design and are influenced by the choice of light source and the space's structural surfaces. Light changes according to the time of day or year and is therefore one of the labile design elements. As a means of spatial design, light needs to be planned so that it is able to adapt to different uses and times of day. Spaces can be open or closed to light depending upon what is required by the use or individual need. They will have different effects according to the type and position of the light openings.

Ω
\\ Note:
Candela (Latin: taper, candle is the unit of measure for light is the): the luminous intensity, photometric base unit for lumen (cd).
Lumen (Latin: light, lamp), the luminous flux, photometric unit of the luminous flux (lm).
Lux: luminous intensity and specific brightness (luminous emittance), measured with a Luxmeter, it can be converted into luminous flux and und luminous intensity (lx).

WARMTH, HUMIDITY, SOUND, SMELL

The space's physical conditions are also a means of spatial design, yet they require a more dynamic approach because they are constantly changing. Spatial conditions are defined by the material qualities chosen for the spatial shell, or the membrane between the interior and the exterior.

Warmth

A material's thermal conductivity determines how quickly heat will be transferred from a warm body or the space's air, and the way in which the exterior and interior temperature difference is balanced. Is the thermal conductivity of the material very different from its surface temperature? Is there an unpleasant movement of air that will make a room feel subjectively cool although it is sufficiently heated? The thermal conductivity of building components that are in regular contact with the user (floors, door handles, seating, and so on) also needs to be considered, in order to create a desire to spend time and feel comfortable in the space (for example, to walk barefoot in a private living room).

> ℘

Room temperature is regulated according to the particular use of the space. Physical work, for example, requires a lower room temperature than office work. Older people often require warmer temperatures than younger people; and the perception of heat or cold varies from one individual to another. Heating systems for interior spaces are categorized either as direct heat (such as the sun or a tubular radiator) that travels directly to the skin, or convection heat that is transferred by air. Room temperature also establishes the room's humidity.

Humidity

Water vapor is absorbed by the spatial shell and room temperature. Humidity is the term used for the percentage of water vapor in a space or in the Earth's mix of atmospheric gases. The percentage of humidity measures the degree of water vapor saturation in the air.

> ◌

Skin is able to sense the level of humidity almost immediately upon entering a room. The relative humidity of a room affects one's sense of

℘

\\ Example:
The level of body warmth in a hand is more quickly transferred to a stone than to a piece of wood. Because the stone cools a hand much more rapidly, the stone feels colder than the wood even if its surface temperature is the same.

◌

\\ Note:
50% relative humidity means that the air contains only half the maximum amount of water vapor it can hold at any certain temperature. At 100% relative humidity, the air is completely saturated with water vapor; if this amount is exceeded the surplus moisture turns into condensation that gathers on the space's surfaces or develops into mist.

Tab.6: Reverberation times according to use (frequency range 100-5000 Hz)			
Recording and sound studios	Classroom, lecture hall	Combi-office	Concerts (depending on the type of music)
< 0.3 sec.	0.6-0.8 sec.	0.35 sec.	1.5-3.0 sec.

wellbeing and consequently one's health. For example, dust bonds together in conditions of high humidity, while low humidity can dry out the sinuses and ultimately cause illness. Moist surfaces in a space often result in mold that can also be detrimental to health.

Sound   Disruptive noise can be controlled by acoustic spatial design. Noise affects the acoustics of a space and is perceived almost immediately. The spatial shell's attributes can function well as sound protection against noise from outside traveling inward or vice versa. The acoustic effect of materials is defined by the degree of sound absorption, which lies between 0 (no absorption) and 1 (complete absorption). The degree of absorption is conditional upon impinging frequencies. There is also a distinction made between two different types of effects: porous soundproofing absorbs sound into the material. Inside the pores, friction transforms the sound energy into heat and thus diminishes the sound reflected by the material. Soundproofing that relies on vibration vibrates with impinging sound waves; this resonance reduces the level of sound reflection.

The most important factor in spatial acoustics is reverberation time. This is the time it takes for a sound to decay in a space, and should be tailored to the space's use. › Tab. 6 For a concert hall, this aspect has to be designed as precisely as possible, yet other spaces such as large offices also need specific sound attributes. Large sacred spaces often have long reverberation times and strong sound reflection; in comparison, very close spaces often seem small and claustrophobic.

Smell   Even the evaporation from material or other people's perspiration can have an impact on the atmosphere one perceives while spending time in a space (library, church, school, locker room, and so on). A strong smell can overpower all the other spatial aspects and make it unpleasant to remain somewhere. Fragrances can produce a pleasant spatial impression if they create positive associations. They are used in certain areas in shopping malls and department stores to create a pleasant experience.

MATERIAL, TEXTURE, ORNAMENT, AND COLOR

> ✎

The effects that the surfaces comprising a space have on the viewer are defined by all of the materials used. Other, fundamental material qualities also have an impact on spatial impression in addition to the specific material qualities or the texture of surfaces and materials. › Tab. 7

The texture of the material is first determined by the way in which the material has been handcrafted or industrially processed, but it is also a result of use, deterioration, or erosion. Most materials can be produced with coarse, fine, smooth, matt, polished, or rough surface textures.

> 𝕠

The surface texture quality also influences the lighting conditions, as well as spatial acoustics, temperature, and indoor humidity.

A room can seem as inviting and pleasant as comfortable clothing if the vibrations of the surfaces of a space are adjusted to the users' movements.

New material

New materials such as nanomaterials and composites are being developed constantly and are used in spatial design. Nanotechnology has made new material surfaces, coatings and textures available that fulfill specific functions. The changes in surface structure are so minute that they cannot be discerned with the human eye.

"Composite" is the term used for the influx of newly developed material combinations that have improved constructive properties. For instance, transparent concrete was invented while experimenting with new mineral aggregates. They are reinforcement with synthetic material instead of steel, which makes it possible to construct much thinner plates.

Fabric and paper

Fabric and paper are temporary spatial design elements, because they deteriorate quickly with use. As with furniture, fabric and textiles can be moved and used in different areas. They are pliable, light in weight, have various textures and are available for design concepts in a broad palette of colors. This makes them very popular as a means

---

✎
\\Tipp:
The different materials intended for a spatial design can be presented in a materials collage that displays the interrelationship between the materials that have been found or planned for the project.

𝕠
\\Note:
The manner in which one walks along a wooden beam floor with floorboards, on screed in a reinforced concrete building, or on sheet steel is very different due to attributes such as elasticity, sound, and sometime heat absorption.

Tab.7:
Materials for spatial design (selection)

Natural stone	Artificial stone	Wood	Glass	Natural fibers and fabrics	Metal	Synthetic materials and composites	Other
Plutonic and effusive rock	bricks	soft wood	industrial glass	_ felt and fleece	cast metal,	straw	mineral binders
_ gabbro	lime sand brick	hard wood (including fruit wood, burl wood)	laminated safety glass (LSG)	_ textiles	rolled metal	adobe	plaster
_ granite	clinker			_ cane	alloy	linoleum	lime plaster
_ diorite	earthenware slabs (cotto)		window glass	animal fiber (spun/ drilled)	iron	asphalt	cement plaster
_ basalt			glass tubes		(steel, stainless steel)		gypsum plaster
_ diabase	adobe	tropical wood		wool (sheep, alpaca, llama, angora, cashmere, camelhair, mohair)	copper	epoxy resin	
_ pumice stone			wired glass		lead		new materials
_ basalt lava	cast stone/ (cement, water, sand/ gravel)	bamboo	fiberglass		nickel	Plexiglas	nano materials (1–100 nm)
_ porphyry		straw			aluminum	acrylic	
_ tufa		reeds	glass stone		zinc	glass	
sedimentary rock		all wood material	glass fiber	hair (goat, cow/yak, horse hair)	tin	foam	painted coatings (acrylic, synthetic resin)
_ slate	_ in-situ concrete	wood fiber	glass mosaic	silk (mulberry, tussah, conch)	titan	rubber	
_ limestone	_ terrazzo	cork	glass pearls	horn	silver	mineral rock wool	fiber materials
_ (lime sand brick	_ ready-mix concrete	coconut shells		fur	gold	film	
_ rimstone)	_ concrete block			leather		sheeting	microfibers
_ conglomerate	_ screed				solid (tubes, rods)	resin (coatings)	
_ dolomite	_ agglo marble			plant fibers	plates	thermoplastics	mineral fibers
_ greywacke	_ transparent concrete			_ linen	wire	_ polyethylene	asbestos
_ sandstone	_ cement lime stone			_ ramie	textiles	_ nylon	
_ quartzite				_ flax	film	_ PET	fibers from natural polymers
				_ coconut	sheeting	_ polystyrene	_ viscose
metamorphous rock	ceramics			_ cotton	alloy	_ polyamide	_ polyamide (nylon)
_ gneiss				_ kapok	foam	_ polyester	_ modal
_ marble	quartz material			_ stinging nettle		_ polypropylene	_ cardboard
_ quartzite				_ hemp			_ paper
_ slate				_ jute		thermosets	_ wallpaper
				_ sisal		_ polyester	_ paper mache
				_ bamboo		_ bakelite	
				_ grass		_ nylon	chemical and ceramic fibers
				_ (grass wallpaper)		_ polyurethane	
						_ synthetic resin	
						_ epoxy resin	
						_ melamine	
						elastomeres	
						_ caoutchouc (rubber)	
						_ polyurethane	

of spatial design for walls, ceilings, floors, furniture, or opacity. Typical textile elements and materials in spatial design include:

_ Carpets (floor and wall): wool, synthetic material, cotton, silk, jute, sisal
_ Wallpaper/tapestries (wall and ceiling): wool, cotton, silk, linen, metal
_ Curtains (interior and exterior wall): cotton, wool, synthetic material, silk, linen, metal yarn
_ Blankets and pillows: (furniture): wool, cotton, silk, linen, synthetic material

The material qualities of textiles and paper resemble clothing more than stone or metal and, consequently, people consider them pleasant and familiar. People like to touch them because they are mostly soft, flexible, and do not absorb body heat.

In spatial design, textiles and fabric also serve to absorb sound because they are porous and have a large surface area. They show the movement of air and provide shading or protection from rain.

Ornament Ornament is the term used for a repetition of abstract or figurative forms or objects (Latin: ornare = to decorate). The pattern this creates can be used to design and structure spatial elements and surfaces. Ornaments can also represent a certain symbolic meaning and have the effect of lettering. Wallpaper is often decorated with an ornamental pattern, but natural stone can also have a striking grain pattern and look decorative. In the ground plan of a city, the arrangements of buildings that appear repetitively can create a pattern. The joints between the surfaces of a sidewalk can also be designed and perceived as ornamentation. > Fig. 70

Many abstract ornamental patterns are based on natural forms or images. An ornament can be painted onto a surface or used as a three-dimensional element to structure a room. In Gothic churches, there are many ornamental structural elements and decorations on the columns

\\ Note:
Islam forbids the use of images, so very fine, complex, and diverse ornaments based on script were developed over time and used in textile and spatial design.

Fig.70:
Ornament in a public space

77

and walls that enhance the building's three-dimensionality by reflecting light and shadow.

Ornaments are often used to structure large surfaces or spaces that would otherwise seem too large, empty, or even threatening. Large surfaces and spaces are divided into areas and sections that correspond to human scale so that users can more easily position themselves in the space. Ornaments fill a space with visual and tactile sensory stimulation and thus lessen or eliminate the impression of emptiness. However, too many differently structured or overly elaborate, three-dimensional looking ornaments can overpower the spatial impression.

Color

Every structural surface in a space reflects natural or artificial light and is perceived as color. The color effect is determined here by the material, its texture, and its surface qualities. The overall color effect of a space varies in intensity, and is conditional upon light, the degree and angle of reflection, as well as the color attributes of the material's surface.

Color can be used to divide spaces into areas and zones, and can emphasize or underplay the focal point of a room. The ceiling in a white room with a black floor will seem higher than if the colors were applied in reverse. Colors also have an influence on how we perceive spatial boundaries: pale, low-contrast colors give spaces the impression of being wider and larger than dark, contrasting colors, because the distance between the viewer and the structural surfaces and dimensions cannot be as clearly read.

Color can enhance the spatial effect of depth if cool-temperature, low-contrast colors are used for the background.

FURNITURE—FIXED AND MOVEABLE ELEMENTS

Generally speaking, the elements of spatial design are all the things that create spaces between their surfaces. There are fixed and moveable

\\ Note:
Since the color effect of the reflected light is only possible if perceived by an individual viewer, it is difficult to express and present this aspect of spatial design in words. Hence, color guides and samples are used to choose and present colors.

\\ Note:
Color can even influence the amount of time a visitor spends in a room. Experience has shown that an intense, orange-yellow-red color scheme reduces this amount of time more than white-green. Fast-food restaurants use this tactic and choose an orange-red-yellow scheme to ensure customers only remain for a short period of time after they have finished eating, and make room for the next guests.

elements of a space. Columns are permanently installed elements of the support structure, but the position of furniture is flexible. Spaces are always influenced by the interaction of all of the elements that form the spatial volume. Even if one element has a stronger effect than the others, the spatial quality is still the sum of all elements together. They form the space between their surfaces and are hence always interrelated. › Fig. 71

Furniture

Furniture is a spatial design element that is either permanently integrated in a spatial shell or can be repositioned. It forms a flexible secondary structure within the primary structure of the spatial shell. Furniture is a means for the user to personalize the space. › Fig. 72

They form independent relationships; a table is given different chairs, and a certain grouping of furniture creates an area within a room. Furniture can make a spatial impression suddenly seem very unfamiliar. Yet on the other hand, furniture can support or even enhance the existing spatial statement and the effect of the space's primary structure.

The size of furniture makes it appealing to use and to touch. It is a very popular design element, because it is well suited to and based on human scale and also creates subsections of space within the day-to-day spatial experience. Furniture often serves as a variable spatial element in public city spaces, apartments, or offices, as it can structure

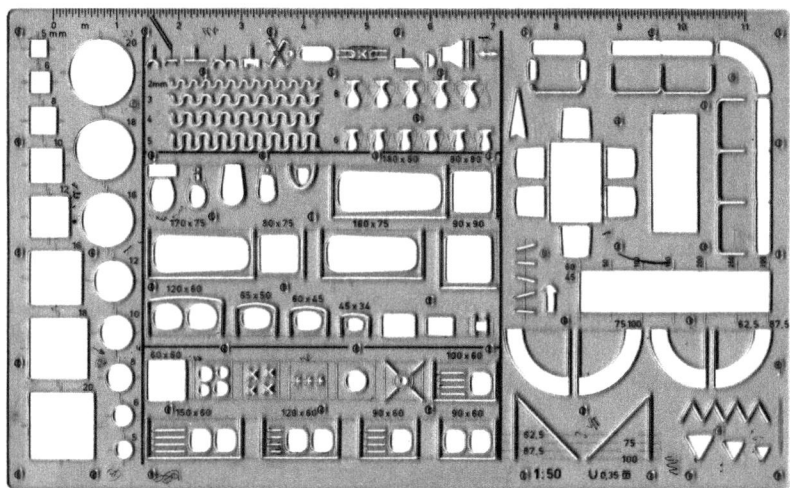

Fig.71:
Furniture template

79

Fig.72:
A variable furniture element can adapt to the given spatial form, whereas here a functional piece of furniture was developed specifically for the smaller space.

the spatial design and can be adapted to different uses. Furniture can also be a tool to diminish or quickly and effectively increase a space's density. Furniture helps people to create a physical connection to the space. A chair in a room has a very inviting effect, simply because it asks the visitor to take a seat.

There is often a piece of furniture in a private room that is associated with memories of family or particular events. This makes its form and attributes less important. The piece of furniture becomes a medium for personal emotions and memories.

> ♀

♀
\\ Example:
The dimensional aspect of furniture differs greatly: a typically small bed can grow to be as large as the room if it is a four-poster, and hence become a comfortable space within a space.

IN CONCLUSION

The objective of spatial design is to give a built space or site a presence that can be perceived with the senses and the cognitive system. It is achieved by implementing functional, technical, intellectual, and aesthetic requirements. Human existence takes place within space and time, and in the process, diverse spatial qualities are constantly being perceived sensorially or with the cognitive system in an intense or casual manner. At the same time, people are always changing their spatial environment, more or less actively and in many different ways. People and spaces are therefore in a constant, dynamic interrelationship. Spaces can be designed so that many aspects of human existence are considered and addressed, and so as to provide users with appealing amenities and various possibilities for interaction.

Spatial design serves to form the built environment with a certain atmosphere that is able to provide and ultimately positively influence the sense of individual wellbeing, social interaction, the way one behaves and acts in a space, and finally, even to facilitate a relationship to the spatial environment in the first place. The mood or spatial atmosphere is the sum of many different sensorially and cognitively perceived spatial phenomena.

Whether a spatial design is accepted by users, visited and further developed throughout time, will depend on whether the complex interrelationship between people and space has been successfully addressed. A space's atmosphere can be designed to be lasting or temporary, and most change over time. Even the simple use of a pre-existing space, without redesigning the raw form with construction or material elements, can change the atmosphere because it is redefined by the presence of new users. Consequently, a space's atmosphere cannot be completely planned in advance. Even the spatial impression of a new building cannot be easily conserved for the duration of its use.

Yet spatial atmosphere can still be designed, which is why this book has introduced a diverse repertoire of spatial design elements and means. The type of spatial atmosphere and the way it is perceived are influenced by the lasting and complex interaction between ideas and a spatial concept, activities, the presence of people, light, spatial form, the choreography of spatial sequences, texture, the materiality of the space's structural elements, and acoustics. All of these factors can be strategically employed to create an insightful spatial design.

One of spatial design's most exciting tasks is influencing the future effect of a space, as well as the consideration and decision about whether or how users can be prescribed either a clear spatial structure, or given the opportunity to personalize their own space.

APPENDIX

ACKNOWLEDGEMENTS

Bert Bielefeld, Dortmund for his patience and critical editing

Tina Jacke, Siegen, for image processing and diagrams

Petra Klein, Siegen, for her organizational support

Sigrun Musa for image processing and diagrams

Judith Raum, Berlin for critical editing

LITERATURE

Rudolf Arnheim: *The Dynamics of Architectural Form*, University of California Press, Berkeley and Los Angeles 1977

Gaston Bachelard: *The Poetics of Space*, Orion, New York 1964

Franz Xaver Baier: *Der Raum*, Walther König, Cologne 1996

Gernot Böhme: *Atmosphäre*, Suhrkamp, Frankfurt am Main 1995

Otto Friedrich Bollnow: *Human Space*, Princeton Architectural Press, New York 2008

Michel de Certeau: *The Practice of Everyday Life*, University of California Press, Berkeley 1988

Fred Fischer: *Der animale Weg*, Artemis, Zurich 1972

Kenneth Frampton, Harry Francis Mallgrave: *Studies in Tectonic Culture*, MIT Press, Cambridge 2001

Walter Gölz, *Dasein und Raum*, Max Niemeyer Verlag, Tübingen 1970

Max Jammer: *Concepts of Space. The History of Theories of Space in Physics*, Harvard University Press, Cambridge 1954

Hugo Kükelhaus: *Unmenschliche Architektur*, Gaia, Cologne 1973

Wolfgang Meisenheimer: *Choreography of the Architectural Space. The Disappearance of Space in Time*, Dongnyok/Walther König, Paju/Cologne 2007

László Moholoy-Nagy: *The New Vision. Fundamentals of Design, Painting, Sculpture, Architecture*, Faber, London 1939

Paul von Naredi-Rainer: *Architektur und Harmonie. Zahl, Maß und Proportion in der abendländischen Baukunst*, DuMont, Cologne 1982

Christian Norberg-Schulz: *Genius Loci. Towards a Phenomenology of Architecture*, Rizzoli, New York 1980

Colin Rowe, Robert Slutzky: *Transparency*, Birkhäuser, Basel 1997

Bernard Rudofsky: *Architecture Without Architects. A Short Introduction to Non-Pedigreed Architecture*, University of New Mexico Press, Albuquerque 1987

PICTURE CREDITS

Images on the following pages are credited to the authors:
2, 3, 6, 8, 9 left, 11 left, 14–18, 21 left and right, 24 right, 26 left and right, 28–30, 32, 33, 35–39, 41, 43, 44 left and right, 45 left and right, 46–49, 53–55, 56 left and right, 58–62, 64 right, 65, 66, 67 left, 68.

THE AUTHORS

Ulrich Exner. Architect and engineer, professor in Spatial Design and Planning in the Department of Architecture and Urban Planning at the University of Siegen, and freelance architect in Frankfurt am Main.

Dietrich Pressel. Architect, research assistant in Spatial Design and Planning in the Department of Architecture and Urban Planning at the University of Siegen, and freelance architect in Frankfurt am Main.

导言

　　空间是人类存在的基础，大多数空间环境都是经过人为设计的。人们日常生活都是在城市、景观、建筑以及室内等空间中发生的。尽管地震或战争等灾难能够瞬间破坏环境，人们依然相信他们所在的人工或自然环境是永恒的。人们往往以一种新鲜的方式，通过个人感觉来直接感知空间。在不同的空间中，人们会进行不同的活动，例如有的空间适合行走、休闲、睡觉或者工作，而有的空间却不适合这些行为。一片森林或一个街道在白天看起来非常吸引人，但夜幕降临后却让人感到危险。只用几秒钟，人们就能感受到一个空间是太近了还是太大了，是安全的还是危险的，是吸引人的还是令人厌恶的，而所有这些感受都会相应地影响我们的行为。徒步者总是根据特定的标准慎重地选择休息地点：有日照、无强风、较凉爽、景色宜人、环境噪声小，总之要足够安静以保证休息不被打扰。但是，这些要素并不能够单独被感知，所以很难详细描述出这样的场所环境。因此，空间场所是由各种要素汇集在一起，场所感也是由这些要素的感知共同形成的。

　　人们根据抵御自然力的需求、不同的行为模式和不同的生活、工作模式，以及他们的愿望和价值观来设计空间环境。但很多空间环境是根据某些人的个人利益或者某政治集团的意愿，由无关的或指定的人来设计的。建成空间能够通过形式、材料、光线或色彩来激发人的感知和思考，其尺度能够给人以安全感或者私密感，其设计能够让人们感到惊喜、惊奇、愉悦或舒适。所以，空间容器的创造同时也是让空间充满活力的过程。空间设计作为一个建设实施过程，可以用定义人类存在的若干参量来进行描述，诸如文化 - 观念的、特定场所的、经济的、政治的、社会的或功能限定的等。这些因素容易产生持续的变化，也总是影响并改变着空间。空间设计应该辨别这些需求和想法，进而为某些个体或某些相关群体服务，其过程可长可短，有时长达千年，有时则只有几个小时。

　　不管是在室内还是在室外，空间设计可定义为活动空间功能的任何类型。空间设计的核心认识是空间作为环境中人、物及与其他要素之间的联系，能够在感觉和认知上获得感知。接下来，我们将讨论建成环境和自然环境的人类感知、空间的典型现象、空间设计要素和方法。

图1:
罐头外表面限定体量，同时表达其表面材料与设计

空间感知

　　空间设计及其效果取决于人类对周边环境的感觉和认知。所有对环境的感觉和刺激经过大脑的处理，反过来又会影响个体的感觉、行为和运动。

　　人类有多达 13 种感觉，包括视觉、听觉、触觉、嗅觉和味觉五种主要感觉以及平衡感。有些人不能拥有所有的感觉，或者不能感知或不能全部感知特定的感觉刺激，例如光线或声音。基于恒定的空间方位，平衡感可以感知重力，进而感知空间的垂直状态。

　　在我们不需要彻底了解空间全部特征的情况下，人类感知引导着每个人的日常活动。我们不断使用新的空间。场所中的诸多信息能够通过感觉与认知系统快速得到处理，在还没有启动思考程序的情况下自动影响人们的行为。人们并不需要认知空间的个别特征，完全可以通过对感知和信息的处理，很快就可以判断出空间是舒适还是不舒适，是封闭恐怖还是开放安全。我们都有过类似的感受，当我们一进入某个餐馆，马上就会知道自己是否喜欢这个空间。

　　空间感知是独立的。多年以后，成年人会感觉度过童年的场所比记忆中变小了很多。同时，很多空间特征是以相似的方式被人们感知的。如果仅对一个人，方向感知系统是不能发挥作用的。对空间环境的感知几乎与人们的活动同时发生，而空间特定的属性则会促使活动的产生。

提示：
　　认知系统这一术语是指与人类感知、学习、记忆和思考相关联的人类机能；换句话说，即人类思考和心理过程。

提示：
　　在希腊，感知即美感。哲学上，美感一词是用来描述感官知觉。但在口语中，美感是美丽的同义词。

图2:
只有通过近距离接触才能感知到材料的具体特征

图3:
受限于人类视域的直接视线

图4:
在摩洛哥沙漠中,估算"通达天堂的台阶"
(Himmelstreppe)的距离

图5:
在行进两个小时后到达"通达天堂的台阶"

P12 **封闭和距离感**

感知主要是人类视觉、听觉、触觉、味觉和嗅觉五类感觉综合的结果。这些感觉的强度因人而异(表1)。这五类感觉只能以整体的方式被人们感知——例如,当人们看到一幅粗糙表面的木版画,就能够感受到其特有的肌理和味道。

封闭感

人们通过嗅觉、触觉和味觉建立与被感知的实物的直接联系。这三种感觉往往不需要光线条件,而且大多是容易获得的。由于通过接触表皮可以感知空间形体,触觉成为感受空间品质的必要方式。

视觉	触觉	听觉	嗅觉	味觉
1000000	1000000	100000	100000	1000

不同大气条件下的能见范围 表2

非常清晰	清晰	轻度多云	多云	乌云，轻雾	阵雪，大雾
50–80km	20–50km	10–20km	4–5km	2km	0.01km

距离感　　　　　　听觉和视觉信号在人们获取感知的过程中共同协作。神经为这些信号建立联系网络，并且提供环境的方向信息。因此，在有特殊声学信号的空间中，比起在声音嘈杂的空间中，视觉更具有选择性。通过眼睛的晶状体，视觉信号在视网膜上形成一个环境的二维图像。同时，对于视觉信号的理解是有条件地建立在个人经验基础之上的。因此，这个二维图像在大脑神经系统和个人经验的共同作用下，最终被感知为一个复杂、混合的空间。

这一点可以从图4和图5看出：对于图示的山脉，其形体在一定距离之外很难被识别，只是看到一个平面形式，人们很难估算出到达的距离。然而，当作为欧洲中部地区典型的场景，人们就能够通过类似的空间和能见度情况（表2）相对准确地估算出距离。

示例：

通过摸、看、闻来感受材料的质量。当这三种感觉都能够被感知并且处于均衡状态时，这种材料就是适宜使用的（见图2）。

　　　　　　认知系统

　　如上所述，人类智力或认知系统或多或少有意识地传达出某种空间感官意象，这种意象影响着人们的行为、思维和情感。一个空间要素作为一个感知或者记忆符号，能够激发人类的本能行为。

作为符号的
空间要素　　　　空间感知的这种方式与阅读文本相似。与语言学的理论和方法相类似，知觉感知的刺激是从作为符号的空间要素中"阅读"出来的，并通过人类智力系统表达和解释这些要素的意义。空间要素因此被认为是数据传达器，能够比要素表层含义传达出更多的内容。

　　　　　　空间现象学

　　现象学的哲学代表理论认为人类感知直接影响空间体验，也就是感官认知能够决定人类行为。人类不需要经过思考，就会有感觉和意识。在人类发展进程中，物质体验已经固化了人类对于事物、空间和时间的观念。自从人类存在，人体就与环境不可分割，空间设计在学习和知识探求中发挥着重要作用。

图6：
伊斯坦布尔清真寺的空间氛围让人印象深刻

氛围

空间氛围对人来说是很重要的，但却很难精确定义或者计量，只能通过分析方法来进行局部判断。它发散的特征使得人们很难去设计、展示和理解。被蜡烛照亮的房间一般会认为是舒适的。蜡烛闪烁的火焰、富有色彩的烛光和空间界面弥漫着的黑暗，并不是造成空间氛围的全部原因。除了这些可视的感官刺激，其他例如蜡油的味道、火焰的温度和偶尔的、安静的嗞嗞声都营造着具有吸引力的环境氛围（图 6）。

图7：
蜗牛壳的形状反映其内部的生命形式

P16   **空间的类型**

　　许多不同的空间形式都会受到相同因素的影响，比如空间的功能、人们的想法、行为模式和需求，或是类似特定基地的条件。在特定的气候、地域和时间的前提下，这些空间形式构成了在不同文化中均有体现的空间原型。因此，建筑形式常常可以表达出居住、生产和宗教等建筑功能，这意味着空间外形和结构设计能够明确地反映出其中正在进行的活动（图7）。

　　下一节中将会介绍一些常见的空间类型，我们可以从空间设计上很容易地推断出它们的功能。但是单凭功能本身并不能决定形式，一些其他的影响因素也将在下一节中详细讨论。

　　尽管人们为了自身不断变化的需求，不断地改变空间并适应它们，空间的很多结构特征依然得以保留。

P16   **功能性的空间**

　　空间的形式一般都会受到其使用功能的影响。每一座已建成的建筑和空间都是人类相互交往，进行交易，举行仪式，举办比赛，观看表演等的场所。这些行为很大程度上决定空间的设计，反过来，空间特征也会影响使用者和使用功能。一个空间可能是某种特定行为的必

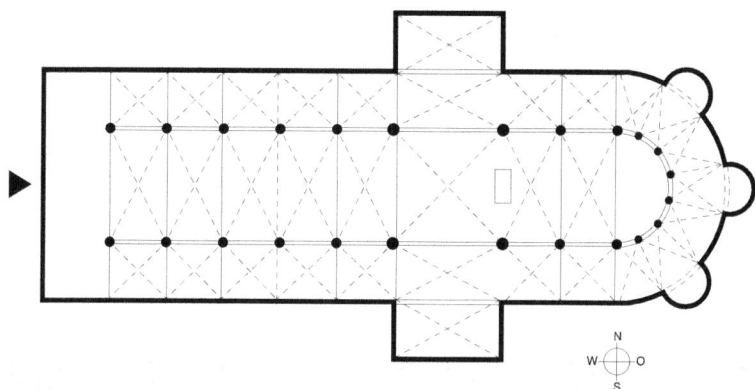

图8:
巴西利卡首层平面

要容器，也可能并不针对某种具体的行为。人们通过对是否能够或怎样清晰地从结构设计中辨别其具体功能的判断，识别和区分空间类型。具体的建筑需求可以很大程度地决定空间设计。如果某种复合形式多次被建造，它往往就会成为一种建筑类型。基础设施和工程建筑的功能往往非常直接地用于某特定功能，且不可能再做他用（图9）。

与巴西利卡空间相反的是一种多功能的空间类型，也是一种空间的功能适应性，同样对空间设计产生影响。因此，尽管一个城市公众广场仅被指定几个完全不同的用途，但单单是它的大小就可以容纳很多不同的活动，比如个人的随意休闲、集体的示威游行、夏日的纪念活动，还有每周的集贸交易等等。

🔍

示例：

巴西利卡是一种精确的空间形式，在整个建造历史中有着不同的变体（来源于世俗建筑）。从西侧进入的长厅（the elongated hall）与朝向东部耶路撒冷的后殿相互对齐。举行宗教仪式的圣坛则被安置在公众能够清晰看到的地方（见图8）。

图9：
不同的技术需求创造出不同的空间类型，例如在阿拉伯国家具有数百年历史、用于通风的传统风塔，或者是在许多工业国家常见的发电厂冷却塔

P18

特定场地的
空间类型

> 🔍

场所精神

特定场地的不同属性会创造出特定的空间形式，因为这些属性在根本上制约着空间的形式和结构。与一块平地相比，一个出挑的悬崖需要特殊的房屋及其支撑框架。

除去这些因素，当地的风力、温度、光照情况也会影响空间排列形式、门窗的数量与类型，或是空间的特色形态（图10）。

图10：
一种由当地条件（穴居）所决定的空间类型

图11：
居住或办公楼大多并不需要特定的场地，几乎可以建在任何地方

与此相比，有很多并不需要特定场地的空间形式，比如机场航站楼。这些空间形式被广泛地应用，并形成多样的文脉和功能上的参照物。其至一些工业化生产的居住和办公楼也相对来说不需要特定地点，可以按照惯例进行建造（图11）。

尺寸　　景观、城市、街道和房间自身不同的尺寸决定了它们的空间形式，同时也决定了它们所能容纳的活动以及它们的主要特征。一个房间只能容纳一定数量的物、人和活动，成为只为少数人服务的私密或者半开放空间。相反，一个城市广场则是一个可以容纳许多人工作、购物、饮食、生活、交流等日常活动并具有适应性的大空间。用巢穴、领土和宇宙来比喻限定人类存在的三种空间尺度：私密的空间、熟悉的环境和公共的区域。

材料　　材料是组成特定场地建筑类型的一项重要因素。利用当地盛产的建筑材料能够建立可复制的典型空间结构，其形式是由这些地方材料的特殊属性和可行性所决定的（图12）。

🔍

示例：

如果开发地段位于带有坡度的街道上，那么建筑入口应该在上层，也需要相应调整首层和空间形式。

95

图12：
取材于当地的东南亚砖土建筑和森林地区木框架建筑

P20   **私密与公共**

空间级别的分类与它们的公共可达性有关。根据它们的功能、尺度和品质等因素，空间具有私密或公共的属性。人们能够很快识别出空间的属性，其行为将受到这个属性的直接影响。因为私密和公共功能之间经常混合或转变，两类空间之间的界限经常处于模糊状态。空间是公共还是私密，是可以通过尺度、社会控制度、可渗透度来进行判断的，具体包括空间界面开口的类型和数量等。

公共空间

公共空间包括建成建筑和社区中所有可能的、开放的空间，往往位于那些一般公众能够以多种方式使用的地点。公共空间可以同时容纳通行、运动、交往、静思等活动，可以接受来自不同社会阶层、国家、文化的个人或团体，他们在这里接触并且交流，不需通过媒介进行交易、表达观点和直接收集信息。公共空间由其规模所塑造。一般来说它应为人们提供充足的活动空间。然而，因为公共空间也是公共交通和运输体系的重要组成部分，汽车、街道和列车同样对其规模产生影响。空间形式设计可以引导和控制人们活动的秩序，因此公共空间也具有政治意义。公共空间设计往往随着时间而被更新或改造，成为诸多功能和意义的见证者。

社会控制

与小的私密空间相比，公共空间中的运动具有更大的自由性。公共空间保障被社会所认同的行为标准，来自他人的社会控制和监督会对公共空间中的活动进行限定和保护（图13）。

图13:
城市公共空间

图14:
巴伦西亚公共广场

　　　　社会控制的缺乏可以使一个空间很快变得荒凉起来，公共活动将不再发生，也不再会吸引人们进入和停留。

广场　　　　公共广场和建筑一般都会被赋予象征意义或地标功能，并且结合其背景可以影响相关地区的城市结构发展。政治、科技、经济和宗教的发展，甚至交通与通信的新技术都可以不断地改变公共空间的设计、意义和功能。公共空间受所在场地的影响。特殊气味、特殊声音、气候条件或者人们在这个特定场地上的着装、行动和活动共同构成这个场地的总体印象，并且决定着空间的使用。由于文化和气候的不同，南方和北方地区公共空间及其设计与使用方式会有很大的差别（图14）。

提示：
　　在一个热闹、有社会控制力的公共空间，如果发生暴力袭击，人们会很自然地干预与援助。但在公共景观空间，这种控制力就会被减弱，这将导致人们产生一种完全不受限制的自由感，也可以认为是恐惧感。

图15:
带有广告牌和立体物的住宅

只要有人类生存的地方，公共空间就会受个人利益和政治集团意愿所左右。使用功能的控制和对空间的设计也是一种权力的体现。

公共空间往往通过明确方向指向及所需的众多符号、要素来进行塑造。

私密空间　　　私密空间是一种能够保护个人隐私的空间类型，在其中发生的活动往往是不能被公众看到的。

提示：

私人经济利益已经使公共空间界面变成了广告和信息的载体。思想和行为的自由度将决定公共空间中能否存在某个特定、独立的部分，如果存在，又将达到什么样的程度（见图15）。

图16：
被均质、一成不变、单一功能的建筑所围绕的公共空间

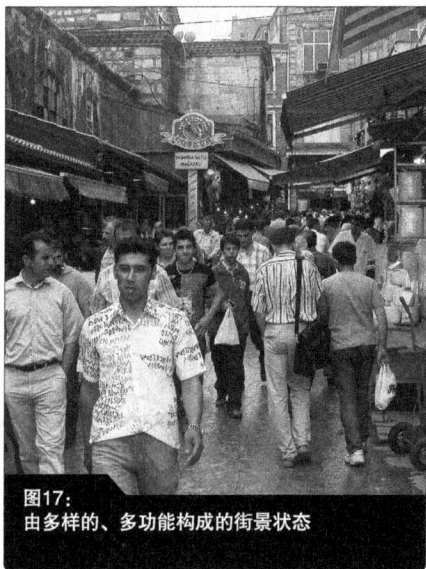

图17：
由多样的、多功能构成的街景状态

　　私人房间或公寓是典型的私密空间。对它们的设计往往是基于人体尺度的，并由人类活动及相关物体所限定，这些活动或相关物体一般不与公众分享。这类空间通常拥有实体的空间外界面，清晰地界定空间的内部和外部，并给人以一种安全、熟悉和亲密的感觉。这类空间的主人可以通过开启门窗来控制外人的进入。

提示：
　　在城市构成中，持续增长的对秩序的需求经常导致单一功能的均质结构，这些空间越来越难被不同的使用者所使用（见图16、图17）。

99

图18:
私人限定的居住空间

居住空间与工作空间

　　居住空间的一个特殊属性是它可以优先依据个人需求来进行设计，但我们仍需明确公共需求和个人需求之间的差别。大多数人共有的需求是基本需求，包括要有安全感、能够遮风挡雨、能够满足日常起居。个人需求比这些基本需求要高，是要在自己所属的空间内能够实现自我发现和自我展示（图18）。

独立性和
隐私

　　居住空间可以反映出居住者的个性。除贴身衣服之外，它距离身体最近。因为这个原因，居住空间中的许多材料和元素之所以被选择是因为它们摸起来很舒适。就像巢穴，它应该给人以亲密感、温暖感和保护感。居住空间又根据功能被划分为不同的区域。视线联系、空间划分、门窗开口都是把空间划分为封闭区域或开放区域的方法。卧室和浴室等满足个人生理需要的房间，基本上是对客人和公众关闭的。它们只有几个外界不容易看到的小窗，并且通常与入口有着较远的距离，而不像客厅，一般位于入口较近的地方。然而，其他房间则应该对朋友和外人开放，并且应该能充分地表达个性。如果居住和工作都在同一屋檐下进行，那就可以将公共空间作为生活空间的一部分，或者相反地，将生活空间作为公共空间的一部分。

特殊居住
空间

　　除了私人空间以外，医院、疗养院、退休老人之家、孤儿院、宾馆等其他场所应满足特殊社会群体的特殊需求。比如说，疾病需要进行空间控制和隔离，从而保证大多数人的身体健康。从另一方面来讲，这些设施也在空间上隔离人体的衰弱与死亡。宾馆为旅行者提供了短暂的栖所，也可以用来居住，和为私人或专业活动提供场地。

工作空间
🔍
›

工作坊、生产间、办公室等空间的设计要满足特定的工作流程和步骤、产品和机器的需求。

工作空间需要有充足的光线、空气和供职员、工人活动的场所，因为这样可以帮助员工们维持一个良好的工作状态。根据员工数量的多少，有相应不同的空间类型。

生产线需要足够大的空间来满足生产需要。它需要根据机器的功能和大小来进行设计，但并不鼓励员工长时间的停留。相反，手工作坊则因为员工在生产中具有更重要的地位，其设计通常会满足熟练工人的行为和需求。工厂店等公众可以进入工作空间的设计一般会优先考虑顾客的需求。

办公场所的设计主要考虑脑力劳动，而较少考虑体力活动的需求。因为脑力劳动对空间的要求大致相同，所以各种办公室的形式一般不会相差太多。

商业企业的目标和功能是赚钱，这就意味着办公场所设计需要考虑成本和运转的产出效益。为了避免对员工身心健康造成负面影响，许多国家都出台了有关办公和空间设计的相关法规，将员工职业健康因素考虑其中，以保障员工的健康和安全。由于办公室和厂房的使用对象是不固定的，所以不大可能赋予它们以个性化。

🔍

示例：
　　印刷厂的支撑结构必须能承受机器的荷载，并保证在印刷振动中建筑的安全性。

文化与休闲空间

文化与休闲空间是为比赛、表演、仪式、购物、展览这些非日常生活、工作的活动提供的空间。它们的设计需要考虑大量的使用者，通常都有很高的举架，这样就可以明显地看出它们的功能。在空间上，它们反映出了人们想要暂时摆脱固有生活状态的一种愿望。它们具有公众可达性，提供多样的空间特性，容纳一般不大可能发生在私人和工作场所的社会活动。城市公园、游泳池、城郊景观都是典型的休闲空间（图19）。

参加宗教仪式和参观教堂、清真寺、寺庙等活动可以满足精神信仰，这种情绪在适当的空间类型中将会得以提升。博物馆、剧场、图书馆是教育的场所，同时也满足沟通和社交的需求。即使商业主题公园或者购物中心作为半开放的公共空间（只针对顾客，并规定开放时间），也能提供一种与日常生活不同的感觉。然而，它们在本质上是受私人经济利益驱使的。

思想空间　可以在某些文化与休闲空间中进行思考。这些空间的设计通常需要满足长时间、集中精力的使用需求，例如学校或者大学。这些思考空间的形成依赖特定的比例、材料、光线和色彩。对每个人来说，

›🔍

这类空间是一种不同寻常的环境氛围的体验（图20）。

宗教空间　宗教空间也是一种思考空间，但它的高高在上的环境氛围让身在其中的每个人都能在无形之中感受到它的魅力。人们很快就可以感受到这个空间的氛围，人们的行为也会随之改变，例如压低说话声音。为了让人们更容易集中注意力并聚焦宗教本身，宗教空间的

›📎

设计会有意激发人们的情绪反应。

🔍
示例：
　　阅览室是用来静思与冥想的场所，其内部一切都只支持一种功能，即集中精力阅读和学习。这些空间提供了临时使用的私密空间，但它们仍然是公众空间，只是严格控制社交活动。

📎
提示：
　　宗教场所的仪式和它特殊的音响效果、不同寻常的空间尺度、往往还有特殊的气味，都可以唤起人们对之前去过的类似空间的记忆。这些因素可以用于不同的功能和目的。

图19：
不同的休闲空间及其不同的设计方法

图20：
适合集中精力阅读的思考空间

图21：
古老的结构和室内场所能够激发思考

图22：
麦加一年一度朝拜克尔白的宗教仪式

运用空间效果去引导情绪，也是一种可以运用在非宗教场所的设计方法，比如政府场所、贵宾接待厅和企业会议室等。

P28

交通与联络

很多空间由水平和垂直的通行及交通流线所界定。组成交通区域的空间类型包括门厅、走廊、楼梯间、地下通道、隧道、桥梁等。根据功能的类型、目的地和速度，存在不同形式、服务于人和交通工具的交通区域。在大多数的情况下，这些交通区域的指向路线非常明确。交通区域的功能限定了城镇或城市中建筑物间的大部分空间。楼梯间、坡道和电梯则控制着垂直方向的运动。

有些交通区域只是完全为了有效地连接不同地点而建立的，而另外一些则同时提供了相当舒适的设施。公共街道通常二者兼有。街道越宽，身处其中的人们就会觉得越舒适。广场和十字路口都是没有方向性的交通区，因为它们能够提供多种路线，也可以提供其他的公共功能。特定功能的交通空间往往会产生剩余的场地，这些

图23:
交通区域/交通枢纽

场地可以用作附属和临时的功能（图 23）。

交通空间　　除了被放在不同地点，充当临时生活和工作场所的汽车、飞机、缆车、轮船以及火车等交通工具，都可以归为交通空间这一类别。这类空间的设计并不是主要考虑场地的具体信息，而应重点考虑其使用功能、前进的方式以及安全性。人们在这些空间里的时间仅仅局限于从起点到终点所需要的时间。但当长途旅行或者堵车的时候，汽车内部空间也会成为人们长时间停留的场所。正是由于这个原因，车内一般都会有柔软的内饰、纺织品、皮革以及电子娱乐设备，给人们提供一个温馨舒适的环境。同时，由于这类交通空间经常会在不同的地点使用或者同时被很多人使用，其相比于固定空间，对使用状况方面的要求则更为重要。有一些提供给交通空间的固定配套设施，比如停车场站、服务设施、火车站的进站与出站区、加油站、轨道电车站、车库、公交车站、飞机场等等。它们的设计更注重交通运输的实用性，但也考虑了这些空间出入方面的需求。

图24：
不同时代用来表达影响力的建筑语言

　　　　影响力

　　公共建筑和私人公寓能够传达出业主对客户、住户对访客的态度，或积极沟通或厌恶排斥。剧院、教堂、市政厅、政治团体总部等空间通过建筑或室内设计手法来传达它们的象征意义。市政厅建筑材料的选择代表着执政党的权威；有着玻璃外廊的政治团体总部隐喻地告诉市民关于"透明"的含义；法院除了它们功能性的平面布局外，还象征着国家的权威性；剧院则为舞台上臆想的世界营造了良好的环境氛围（图 24）。

　　一般来讲，宏大的建筑物可以象征权力，如果在远处有较小的附属建筑加以衬托的建筑也同样可以达到这样的效果。

　　世俗建筑对于权力和影响力的暗示并不像公共建筑那样明显。很多空间的设计是规划师们针对特定人群提出的想法，而不是直接针对个别使用者。而且，经济因素和财政预算通常不会允许住宅环境个性化，这就导致了千篇一律的住宅形式。在规划者的意识里，

图25：
土耳其的苏美拉修道院，一座已经使用了几百年的永久性建筑；和一个用于临时使用的帐篷

使用者只是一个抽象单位。在规划者的决策因素中，不同居民的生活方式是次要的。规划者表达自己的观点并将其强加给第三方。与此相反，居民在自己家中会用个性化的空间配置来满足他们自我表达的需要。精选的家具和尽可能个性化的室内设计都是表达个性的手段，比如特别的窗帘或者一个与众不同的入口门厅（详见"空间设计的要素与方法"一章）。

空间代表了各种秩序、力量、控制、权力的哲学。监狱、封闭的精神病院、有时甚至是整个国家，都是限制或控制居民自由行动的场所。

永久使用与临时使用

空间可以根据使用的持久性来进行区分。因为它们从建设伊始就开始影响空间类型。坚固的形式、耐用的材料和结实的构造可以用来设计永久性结构，例如纪念碑、弹药库、陵墓等。建筑材料的磨损和各种后续的附加设计可以改变原本的形式，让它几乎无法被认出来。这意味着长期使用能够对空间设计产生很大的影响（图25）。

与此相反，产品展销大厅只是为了预设的产品销售期而建设的，帐篷搭建起来只能使用几小时或几天。空间为了聚会被临时性装饰和使用，街道临时成为队列行走和游行的场所。城镇中的闲置场所

107

图26：
破旧的船舶码头被赋予了新的临时功能

图27：
德国埃森关税同盟煤矿工业建筑群，废弃的工业用地临时用作露天游泳池

图28：
法兰克福2006年世界杯期间的公共景观

提示：
　　当一个大公司关闭或者搬迁的时候，它通常会对城市整个区域产生负面影响。闲置场地的临时功能可以衍生出对现有空间的再设计（见图 27 ）。

图29:
一个悬挂的顶棚塑造了一个高30厘米的虚空间

图30:
在悬挂的顶棚旁边添加一些人物，则可以塑造出一个戏剧空间。感知是建立在熟悉的人体尺度基础之上的

会被用作临时使用或者重建来适应新的需求（图26）。

有时，新的临时功能的"植入"可以激活其他建筑，并且对它们各自所在环境产生滚雪球般的影响（图28）。如果能够证明新植入功能是成功的，那么原本的临时结构就会转变成永久性的。

舞台与虚幻空间

与文化和休闲空间相似，舞台空间和布景的设计也创造出与日常的临时性关联。虚幻空间很大程度上是一个空间类型的临时改变。与设计剧场相似，根据舞台设计出若干布景，安装后又被拆卸。观众们在这些布景的引导下进入演出期间的虚幻世界。空间道具与灯光效果除了能够营造空间，还会激发观众的想象力（图29、图30）。

空间想法可以通过舞台背景的透视效果表现出来。把贸易集会和展览的结构设计成舞台透视的效果，这样商品就能够以视觉最优的方式获得展示。

视觉陷阱
虚幻空间

视觉陷阱效果是一种用来打破现实空间边界的透视元素（图31）。

在图纸上或者用三维模型表达的建筑概念是空间设计的构思方案。设计最初以概念的形式出现，但并没有实现。许多建筑师天马行空地产生灵感，但为此却忽略了一些必要的法则，比如重力或者户外天气。这样的工作方式有时会产生不切实际或是功能不甚完善的空间，这种空间更像是一种幻想，而不是实际的功能性结构（图32）。

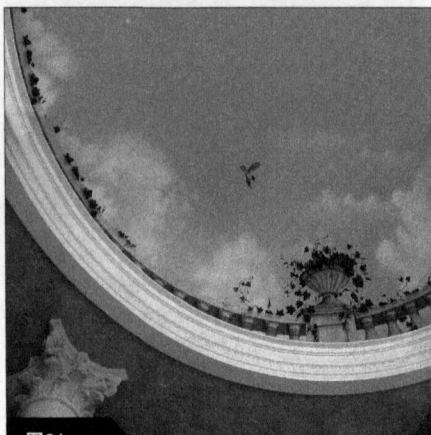

图31：
视觉陷阱效果：真实与虚幻空间

图32：
一个幻想出来的虚幻空间

空间设计的参量

　　除使用功能以外，一个空间也拥有其他与众不同的特征。这些特征能够作为特定场地品质加以强调，它们对空间设计来说具有重要意义。在建筑设计和城市规划中的一个基本设计元素就是建筑或构筑物间的空的空间，能够通过对单个建筑要素和构件的设置来对这些空间加以设计。人类通过感官感觉和心理认知来理解空间。空间现象决定了不同的空间尺度，进而决定不同空间设计方法的类型、应用及效果。这些方法将会在这一章节后半部提到。

环境背景中的建筑

　　每一个场地都有它自己独特的空间环境。一个建筑的设计可以改变它周边空间的形式，反之，周边空间也对可能的建筑设计做了限定。许多复杂、多样的因素对场地产生影响。除了自然和建成环境以外，还有众多可以作为基地环境背景的历史、文化和社会等影响因素。如果与环境特征没有或者只有很薄弱的联系时，参照物的类型、秩序、数量和密度赋予空间设计以文脉性和自主性。

　　村庄、城市和景观是不同的建筑环境背景，它们都能决定建筑设计的类型。比如，我们通常会依据自然采光能力来设定相邻建筑各层的标高和首层平面的方案。

规模与空间尺度

　　空间与建筑的规模，特别是大多数常见尺度，主要是由结构使用方式决定的，通常会与人体尺度、相邻空间有关。我们通过比例来感觉空间大小。当把一个符合人体尺度的较小建筑放在一个大建筑旁边时，会感觉它比真实要小一些。反之亦然。观察者的个人经历，和他所熟悉的、用作参照的特定空间尺度，也对规模感知起到重要的作用（图 33）。

　　在农村平房中长大的人与在城市摩天楼里长大的人，对空间规模的感知是不一样的。

图33:
采石场中的比例关系

　　空间大小效果的相对感知会对人们产生影响。例如，它可以影响我们在空间中移动的路线和方式，或者可以判断这个空间能否给我们带来安全感和私密感。当房间超过特定规模时，它看起来将不再受到限定，与人的关系也不再紧密，甚至已经感觉不到它的大小了。

　　在建筑历史进程中，建筑师与规划师提出了几种基于人体尺度的设计理论系统，包括勒·柯布西耶提出的"模度（Modulor）"理论（详见"空间设计的要素与方法"一章）。

P36

内部与外部

　　每个空间边界都会对"此时此地"进行限定。当附加的空间边界对建筑结构加以界定，人们能够感受到空间深度，这时便产生了内部空间和外部空间。空间外界面成为内外部联系的过渡媒介，开口的类型和数量决定内外部空间之间的关系（图34）。

提示：
　　尺度是相对的，也就意味着判断有可能出错。比如，在大的家具商店里，一件家具第一眼看上去小巧精致，但当把它摆在小公寓里，会觉得它的尺度完全不对，突然变大了。

示例：
　　一个外部无窗的大礼堂会使长时间听报告的观众感到非常的劳累。临街的一个开窗会减少空间的封闭感，同时让观众感到舒服一些，也能够让眼睛和大脑得到一定的放松。

图34：
作为内外部空间过渡媒介的玻璃幕墙

从外部可以清楚地看到，也可以完全看不到建筑内部。玻璃幕墙成为内外部空间几乎无缝的过渡媒介。内外部之间的空间边界可以像玻璃一样轻薄，也可以像中世纪城堡外墙一样厚重，甚至可以置入一些小房间。

开放与封闭

开口的类型与数量，或者空间边界的整体可渗透性，决定了这个空间是开放的还是封闭的。通过开口可以看到邻近的房间或者走廊，它们也可以营造出开放和封闭的感觉。

P37

秩序与偶然

现有景观可以被认为是依据自然条件影响并经过一定组织的空间，但这种景观通常是无序和混乱的。人们通过分区及描绘的方式来组织现有空间，而景观的地形和植被是依据自然法则而形成的，因此任何建筑和城市规划项目都是自然和人工秩序的混合与叠加。

偶然

由于空间不同元素之间有着相当多样和复杂的联系，除了明确目标和计划好的空间设计以外，大多空间设计是留给未来可能性的，或者由使用者自行组织的（图35）。因此，应对那些使用者可以决定用途的区域与已依据预定构成法则完成设计的空间进行统一设计。

当在城市环境中进行建设的时候，常常有一些不同历史时期建造的经典建筑。这些建筑代表了早期的秩序原则。即使历经数百年，人们依然能在城市肌理中辨认出这些建筑秩序感来。新旧秩序和偶

113

图35：
没有遵照规划原则和秩序的建筑

然性建筑在许多城市中出现并交织在一起，很难将它们区别开来，从而带给人们迷宫般、缺乏秩序的感受。

在日常生活中，人们不得不与空间秩序的不同体系打交道。许多体系是由城市管理部门设计的，或者由建筑师、城市规划师和工程师设计的，而并不是由使用者自己设计的。

私人住所是为数不多的可以局部根据自己意愿进行设计的空间之一。私人住所可以传递出这样一种信息：在没有必须遵守的空间秩序原则的情况下，一个人会怎样设计他或她的私人空间。在这里，我们可以看到所谓"囤积"的个性化组织原则。住户把他们的个人物品添加到现有空间，同时调整空间使其能满足个人需求。这些改变很快带来与原有秩序自主体系的差异，哪怕是最细微的偏移都会给人一种混乱的感觉。

示例：
在一些城市中，新建筑建在已废弃的罗马半露天剧院的屋顶上，同时赋予住宅等并非历史性的其他功能。只有在城市平面图中才可以看到古老的空间秩序，也还可以看出新的建筑秩序叠加在老的秩序之上。

图36:
以地平线作为边界的空间深度

方向和导向

　　另一个典型的空间原则是空间队列，这对方向感塑造是必要的。由于重力的原因，垂直方向是上下划分的，而水平轴线则是由视野里左右方向可见的水平线所限定的。房间里的所有元素，连同光线照度，共同决定了我们视觉感知的空间深度范围。然而，活动是对深度加以限定的第三空间维度，使得空间成为有形的体验（详见"空间设计的元素"一章中"时间与空间"一节）。

　　深度是人们感知空间的基础。如果没有它，人类活动将无法进行。地平线是在空间深度上恒定的水平空间边界。因为地平线遥不可及而又无法触碰，人们认为它是无穷无尽的（图36）。

　　排列整齐的空间引导人们沿着空间的主轴线方向移动，进而全方位感受空间。人们通过感官感觉和认知系统来感知空间队列，能够沿着这个轴线走到头。不整齐的空间，比如内院或者城市广场，并不能给人们带来特别的方向感。可是，只要空间足够大且宽敞明亮，人们就有可能在这个空间中逗留。空间导向要求空间的边界必须是有效的。因此，空间的开口越多，其导向性就越不明显。

P39　　　**密度－虚空**

　　一个空间里物品放置状况可以让它看上去是热闹、开放的，也可以是密闭、幽暗的。在空间中无法动弹的感觉很令人恐惧。但是，一个能够自由移动的、空旷的大房间也能够让人感觉到危险。因为在这种情况下，人们没有能够估测距离的参照物，进而不知所措。

图37：
城市密度是由丰富、多种多样的设计以及无数
人类活动形成的

因此，如果不能与标准参照的人或物建立关联，空间的尺度就不会
明晰。有几种方法能够建立感官、人体或大脑之间的关系。一定程
度的空间密度是有必要的，能够给人们提供舒适感。

示例：

　　第一个到达私人派对的客人一般都会逗留
在厨房——一个堆满了各种物品的小房间，很
多活动都可能发生其中，包括客人的活动。厨
房一般比原本用来接待客人的、宽敞的起居室
更具有吸引力。

提示：

　　房间角落里墙面剥落下来的碎片，或者是鲜
艳的黄色墙纸都暗示着使用时间和使用者本人，
它们会给人们带来空间的真实感。

116

图38：
由于材料老化而产生的时间印记

　　空间究竟是拥挤还是空旷的感知是因人而异的，这是与每个人的身体尺度、经历、情绪和运动能力有关的。如果空间中有很多的历史要素，也可以给予人们充实感抑或空虚感。空间密集程度能够快速、直接地表达出空间是否宜人，但却无法定量。我们只可以通过个人经历、文化影响、生理和心理上的行动能力来对它进行评价（图37）。

　　时间与空间

　　空间体验总是与时间联系在一起的，人们总是在不同的时间到达不同的地点。在下一秒钟，即便是相同场地都会因为光线变化、人们注意力转移或是房间物品移动等原因而发生变化。人们步行穿越空间时，同时感受到时间与空间，这是因为步行速度或时间要求决定了这个过程。由于时间和空间同为人类存在的决定性因素，特定的空间记载着某种记忆，特定记忆也联系着某个空间。

　　时间会改变空间材料的物理性质。随着时间的推移，材料状况会发生变化，其原因可能是日照、机械磨损，或者仅仅是日常损耗等（图38）。

　　空间是时光的见证，通常会由许多不同时期的要素构成。因此，空间状况是逝去时光的可视符号（详见"空间的类型"一章中"永久使用与暂时使用"一节）。

空间状态

　　特定空间效果是由温度、湿度、音效、光线和气味等一些物理
化学指标所决定的。所有这些指标都是典型的空间属性，它们共同
合作并且随时间而变化。最重要的是，它们都是在封闭空间中获得
感知的。空间外界面的特性是内外部空间或多或少相互渗透的载体，
对来访者产生影响，例如通过它可以调节温度。与此相似，人类皮
肤也是身体与外界环境之间的过渡媒介，甚至能够感受到最细微的
温度或湿度变化。

　　房间温度对使用者产生直接影响，人们可以根据体温及活动预
先设定房间温度。例如，在低于18℃的环境里，办公效率会大大降低，
但是在这个温度下进行体力劳动则会感觉很舒适。高温甚至会使一
些体力劳动根本无法进行。衣服就如同附加的一层皮肤，也会影响
空间对人体产生的效果。

　　房间湿度与房间温度有直接联系。热空气比冷空气能够吸收更
多水分。当达到露点之后，水蒸气会发生冷凝，在温度低于露点的
房间表面上形成小水滴。

　　空间边界表皮凭借特殊的表面材料反射或者吸收声波，对房间
的声环境产生显著的影响。吸声墙不发生声波发射，声波将渗入墙
体材料，进而被吸收。

提示：

　　当房间表面温度低于空间温度和人体温度，
会让人感到不舒服。因为房间表面会发生恒温交
换和温差补偿，它会从人体吸收热量，使人感到
寒冷。

提示：

　　坚硬的、不具有渗透性的界面可以强烈地反
射声波。但因为回响时间过长，使置身其中的人
很难准确地听到声音，甚至根本听不清。声波多
次反射，很难被吸收。如果音量过大，就会超过
人类耳朵的疼痛临界值，除非安装消声装置，否
则根本无法待在这个房间里。

图39:
由材料决定的空间开口

光

自然或人造光源是空间设计的一个必备要素，也是空间维度和品质的基本信息源。如果空间表面不能反射足够的光线，空间本身将是不清晰的。进一步说，只有存在能够发生反射的界面时，光线才能被感觉到。空间表面反射不同强度的入射光，同时提供空间各个维度的信息。可以通过反射蓝色光谱末端的光线，和保持与最小值对比的方式，来加大空间深度。

P43

材料

材料显示出空间结构的选择，也因此影响着空间形式。材料也决定我们如何去看待工艺的特殊细节、长度、宽度、跨度，以及空间边界类型与特性。因此，空间开口的尺寸和无柱支撑的顶棚跨度都取决于建筑结构是否坚固，或者是否是由木材、钢材和混凝土组成的框架结构（图39）。

正如上文所说，一种特殊材料的质感、成分、色彩和气味，都能影响空间的表现形式与感觉。与此同时，结构要素的可行尺寸和构成方式，以及产生的空间形式本身，都取决于所选择的材料，也取决于材料在工艺和产品方面的特定选择。

氛围

　　舒适感、安逸感和幸福感等都是不能被直接定量的空间效果，但是可以瞬间被感受到。氛围是一个典型、真实的空间现象。空间氛围可以不通过理性理解，而以非常直接并且复杂的方式满足人们全部的感受。"幸福"是一种非常难以定义的效果，部分原因是感觉是主观的。除了功能与精神需求以外，只要有人的地方，氛围就显得尤为重要。它是由人们的活动、构成空间的所有参量，以及通过物质和精神感知到的品质所决定的。

空间设计的要素和方法

下面介绍空间设计可以运用的要素与方法、操作方式及其预期效果。正如上文所述，空间是为满足功能、美学和技术要求而创造的。它是多样空间现象和活动的总和，也是他们持续、综合的相互关联的总和，我们能够通过感受和认知来感知这些要素。空间氛围可以是有吸引力的，但也可以是有压迫感的，它可以引发特定的活动，唤起逝去的记忆，也可以使来访者愿意在此逗留。空间的基本性质对于空间愉悦感、空间设计类型，和人们在空间中的活动方式来讲，都是必不可少的。

想法与概念

设计想法源于所有空间塑造可运用的方法和要素。它可以是尝试性地，亦可通过分析已知变量，还可以凭直觉获得。但一般来说，想法是这三个途径相互配合产生的（图41）。

设计构思是一种对空间的概念，阐述如何满足使用者对空间的需求，以及如何以建筑语言进行表达。它是设计得以实现的基础。

概念 功能和美学的观念影响着人们对建筑和城市的理解。概念是对作为空间设计基础的秩序的选取原则，可以通过一个不同寻常的方案、一个特殊的支撑结构、一种根据环境而建的轴线关系、一些使用者活动的典型轨迹，以及一种对空间及其功能的特定安排来加以表现（图42）。

图40：
城市中兴建了大量建筑，其中大多缺乏对空间设计的考虑

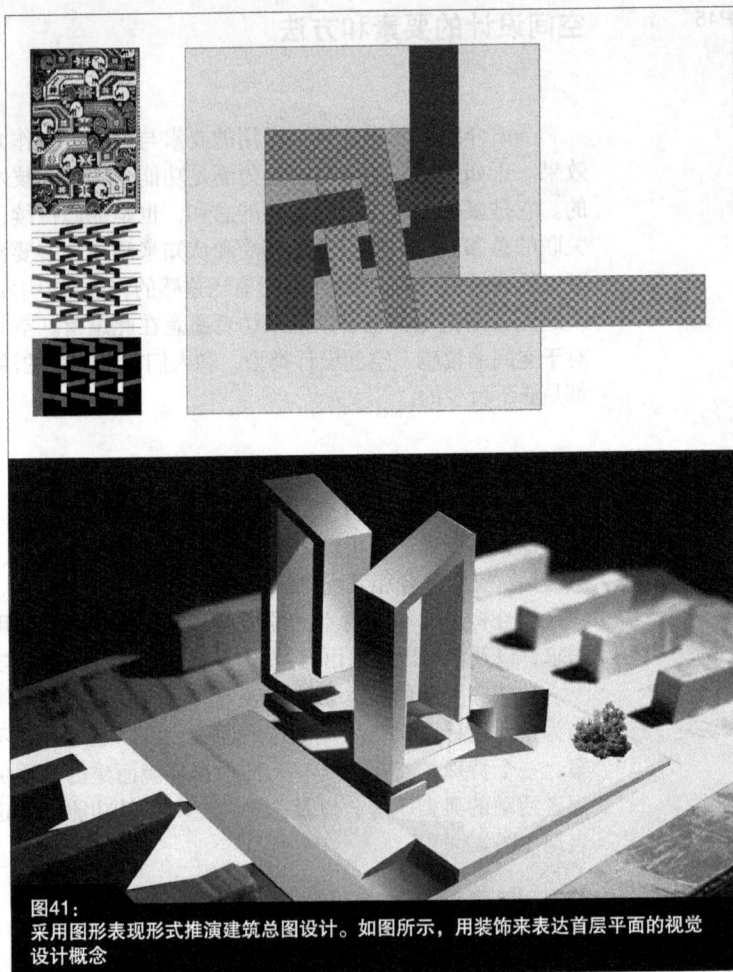

图41：
采用图形表现形式推演建筑总图设计。如图所示，用装饰来表达首层平面的视觉
设计概念

　　设计伊始，往往是凭直觉的或是试验性的，接下来就需要用手绘草图和模型来验证设计能否满足所有的要素需求，能否解决空间中的所有问题，进而进行后续的相应改进，并最终得到最佳的解决方案。

　　许多塑造空间的方法和元素都是可变量，比如色彩、光线、声音、质感等，因此应该有一个主导的设计概念。这个概念生成设计并使得人们能够清楚地理解空间构思。设计概念是空间设计持续表达的手段。特别是对空间结构来说，即便该空间在日后被用作其他用途，人们仍可以看得出这个概念（图43）。

图42：
建筑中建筑（a house within a house）的概念：用新建筑重构原有建筑

图43：
空间概念：一个城市广场表现为一个由直线、平面和体量构成的三维结构。空间体量也支持由若干小片段组成的广场建筑，并增强这些建筑的形体感

提示：

    概念产生的其他信息与想法可以参考本套丛书中的贝尔特·比勒费尔德（Bert Bielefeld）与塞巴斯蒂安·埃尔库里（Sebastian EL Khouli）编著、张路峰翻译的《设计概念》一书（中国建筑工业出版社，北京，2011年，征订号20273），以及卡里·约尔马卡（Kari Jormakka）编著、王昉雯翻译的《设计方法》一书（中国建筑工业出版社，北京，2011年，征订号20277）。

图44：
现状环境被临时功能所改变。如图所示，五年前在空地上建设的一个小型轻型结构建筑，已经成为小型商场

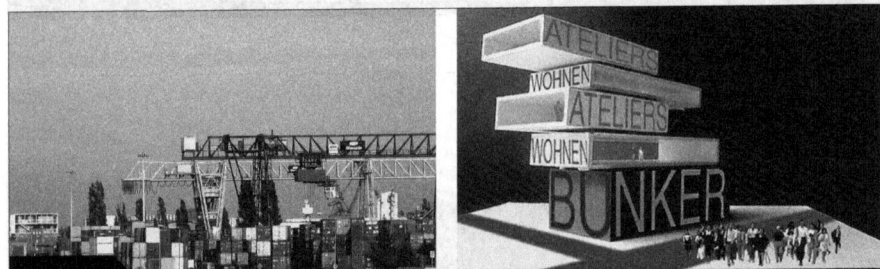

图45：
新的空间结构适用于集装箱码头的使用情况，其设计主题体现了堆叠的原则

功能 　　　如上文所述，居住、办公和工业生产等不同的功能可以塑造不同的空间类型和概念。在现有结构条件下，往往已经有某种特定功能，但对这个功能的重新定义并进行改造可以成为空间设计的一种手法。如果对原有空间内启动一个全新的、不同以往的空间方案，不同的使用人群及他们的活动也将相应地改变空间设计（图44）。

　　　当现有场地被外人占据的时候，空间效果也会被改变。这些人和之前的使用者有着不同的衣着，代表不同的时期，或者有着不同的行动方式。他们带来的一切都会对空间产生影响，会对空间临时重塑，使其适应他们的特殊需求。一种不同以往的新功能也可以用来带来一些与传统的，或不再适用的当代功能相关的重要问题。

环境背景 　　　环境背景是由自然环境或现状建筑属性塑造的。以场地具体环境质量或特征作为参照，对空间结构进行设计（图45）。

124

城市、历史或社会状况可以当作是设计的参考背景。某特定环境的现有功能和使用状况，例如商店的分布、闹市的嘈杂，都会影响空间方案的制定，进而影响新的结构设计。

　　在整体空间关系中，能够或如何表现一个建筑的文脉性，也是一种设计手法。文脉性可以被一堵封闭、坚固的墙所遮挡，也可以有意地利用开窗方式进行借景。通过这种方式，海景窗成为联系外部景观的媒介。如果没有这个媒介，将很难了解所处的环境内容，因此也无法建立与环境之间的关系。

　　从建筑形式、建筑材料质量以及建筑空间设计中，我们可以获取环境的信息。采用当地建筑材料也是一种与环境相关联的方法。使用这些当地材料，建筑的色彩和材质可以与它相邻的建筑相似，进而实现与环境的融合。除此之外，建筑尺度和建筑形式也决定了该建筑是否与环境相协调，抑或是与环境形成对比。

P49　　**空间标记**

　　每次空间设计首先需要了解设计基地和现状空间状况的基本信息。在场地调研时，需要调查并记录地形和现状建筑属性，这一过程被称之为空间标记。通过速写、笔记或文字记录、照片、影像资料、实测等多种形式对空间进行调查，获得定性和定量数据。运用测量设备，尽可能客观地进行计算，以及记录空间的范围和具体场地的特征，进而利用这些数据，运用可行的设计手段制定设计策略。除了卷尺、折尺、水平仪等传统测量工具以外，数字激光测量设备能够实现在三维尺度空间范围的精确扫描。

✎

提醒：
　　因为折尺或者其他测量工具并不是随时都有。在没有测量工具的时候，我们可以参照步伐长度和手掌宽度对距离进行粗略的估算。

📎

提示：
　　为了适应设计任务，通常会调整计算与测量的精度。建造家具比建造街道交叉口需要更精细的单位。这些单位在规划中同样是客观真实的，但通常只是概算。

不同空间尺度的通用比例：图例中的 1cm= 实际的 Xcm　　表 3

景观	城市	建筑	施工图	家具	构造细节
1：100000– 1：2000	1：10000– 1：500	1：500– 1：100	1：50– 1：20	1：20–1：1	1：10–1：1

　　空间标记和技术性数据说明场地的空间属性，也表达出使用者、施工工人以及其他参与建设的人对空间设计的想法。人们往往只是在现场独立、直接地观察空间，因此，设计师采用空间标记方法来表现空间，并传达出场地之外的信息。这样，其他人也能全面地了解空间设计的想法。

全尺寸调查

　　作为空间设计的工作基础，空间通常以缩小的比例进行表达。选择的比例应使空间表达较容易操作（比如能够打印在一张纸上），但也要足够大，能够完整表达出必要的设计细节（表 3）。

　　图纸上所有线条都是现实空间的抽象表达，因为它们只对现实中上下叠加或左右相邻的表面和材料加以表现。空间一般都会以二维形式呈现，比如平面图、剖面图、立面图。这些空间界面表面的正交投影展示、限定和联系着空间（图 46）。

　　正交投影可以表达空间的形体和尺寸细节。但它们并不表达所有的信息，是对复杂性的抽象和简化。其他空间属性的分析与标记有助于形成空间设计概念，例如噪声吸收情况、空间界面材料的质量（肌理、色彩、材料等）、不同时间的场地状况和历史信息等。对于这些类似的空间要素，通常有几种可用的标记方法，例如用仪器测量光照强度与噪声水平。对使用者和邻居的访谈也经常能够获取一些有用信息，而这些信息之前并没有被考虑。

空间位置

　　可以通过视觉形象系列的影像，也可以通过可触碰的三维模型来审视和传达空间设计概念。外行人一般不会理解正面投影，他们更加熟悉透视图和微缩模型，这些更直观的表达方式有助于他们理解空间设计概念。

　　透视图可以准确并写实地表达出真实的三维空间，也可以只是抽象的线框表达。三维空间投射在屏幕上，所有等深线都以斜线的形式延伸到水平线上的一个或多个透视灭点（图 47）。

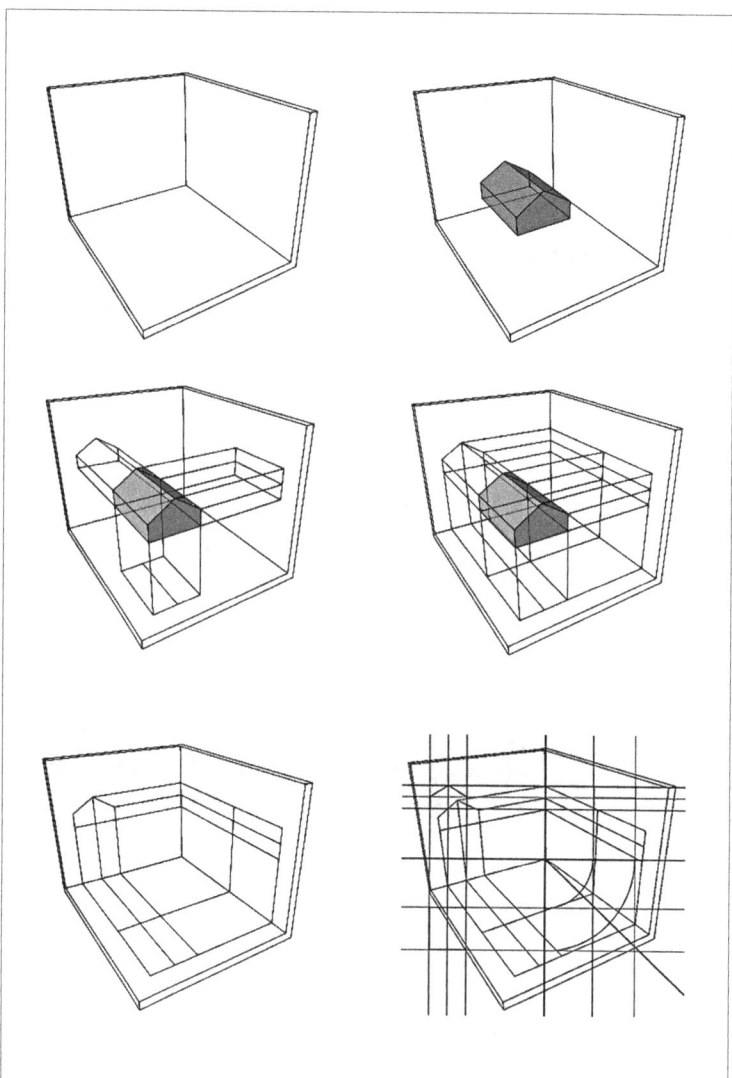

127

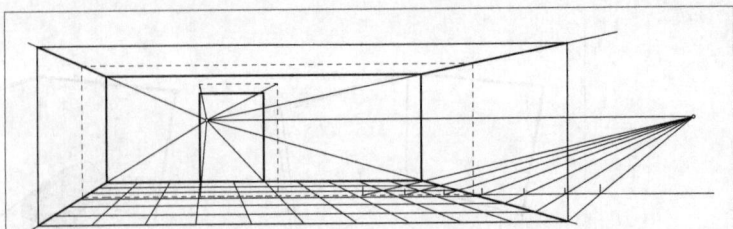

图47：
简单的线框透视表现的空间深度

图48：
将建筑图像投影到模型上，这种空间展示可以评估空间的使用情况

因为每个观察者都有想象和归纳的能力，一个过于精确写实的表现形式可能会让客户和使用者以为设计已经完成，他们将不再对概念提出想法。而手绘图只是阶段性的成果，仍有着可调整的余地，也能够很好地表达设计概念。

现在，空间可以通过电脑模拟来表达三维空间的运动形式。通过调整观察者位置和视高，再现他们每天的空间体验。

空间模型

除了透视表现形式以外，模型也是一种很常见的工具，能够以较小比例模拟和展示空间。模型有助于帮助我们理解空间的联系和尺度，同时也有助于设计和表达空间概念。因为模型与我们日常空间体验最为接近，人们通过模型可以更充分地理解空间并且更直接

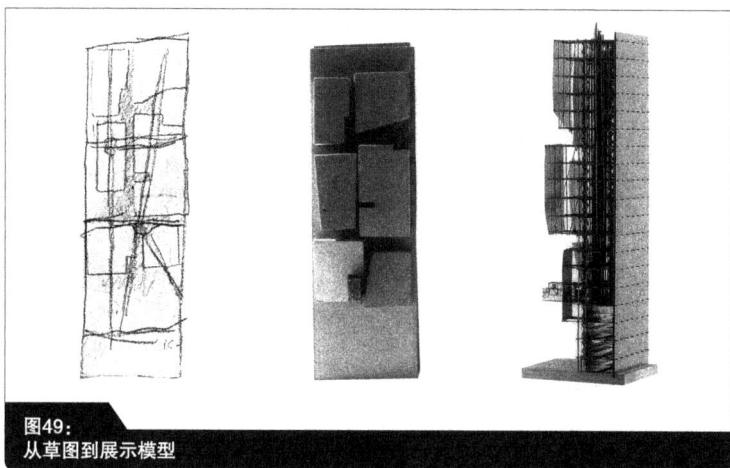

图49:
从草图到展示模型

地理解设计概念。观察者观察模型的位置和视高可以自由地调整。除了视觉方面，人们还可以通过触碰模型感受各种材质，或者用来调试不同的光线效果（图48）。

有些空间及其空间元素的模型，比如立面元素，可以按照1∶1的比例进行建造。通过这个模型，人们就可以探讨方案结构是否可行（图49）。

提示：

有关空间表现形式及与空间概念之间的联系，可以参考本套丛书中的贝尔特·比勒费尔德（Bert Bielefeld）和伊莎贝拉·斯科博（Isabella Skiba）编著，吴寒亮、何玮珂翻译的《工程制图》一书（中国建筑工业出版社，北京，2011年，征订号20282）；扬·凯博斯（Jan Krebs）编著，杨雷、刘婷婷翻译的《计算机辅助设计》一书（中国建筑工业出版社，北京，2011年，征订号20272）；亚历山大·谢林（Alexander Schilling）编著，王又佳、金秋野翻译的《建筑模型》一书（中国建筑工业出版社，北京，2011年，征订号20274）；迈克尔·海因里希（Michael Heinrich）编著，吕晓刚翻译的《建筑摄影》一书（中国建筑工业出版社，北京，2011年，征订号20197）。

构成、比例、尺寸

所有空间设计的要素和方法一起建构空间。这个建构是设计师对空间要素策略性组织和安排的结果。与音乐编曲的方式相类似，设置主调要素和空间，并将它们联系起来。根据功能和美学需要，对单一空间或空间序列来进行组合建构。要素间的功能要求对空间构成产生影响，这就像为特定设备提供施工场地和运行场所一样。建筑和空间构成遵循几何法则、比例关系、本能习惯、二维意象、环境节点间的轴线关系，甚至特定地形产生的景观。

秩序和偶然

大多数空间设计和组成是通过使用者在空间中重新定位的活动和参照物，在日常基础上实现的。甚至一个人的气味和声音都能够极大地改变对空间的印象，在周边不停移动的要素也将不断地改变原有空间的秩序和构成。而且，尤其那些长期使用的结构，结构性要素的具体情况基本不可能完全进行预测。由于我们不可能完全控制设计的方方面面，一些构成部分应该保留其他改变的可能性。正是由于这个原因，许多空间构成整合了那些没有特定用途的部分。

实验性设计

除理性分析方法以外，实验性方法也能创造出精彩的空间。只需要用几个特定的步骤，简单的几何形体就能很快地转化成空间复杂形体。二维矩形和三维规则几何形体（比如立方体）都可以被划分、复制、折叠或根据其他几何法则变形，并最终形成多样的空间形式（图50~图52）。

示例：
在建筑行业里，未来的业主是不确定的，所以一定要给壁橱、床、椅子和其他家具留出空间。类似这些家具设计的未来空间设计已经不再受到"空间编排"的约束。

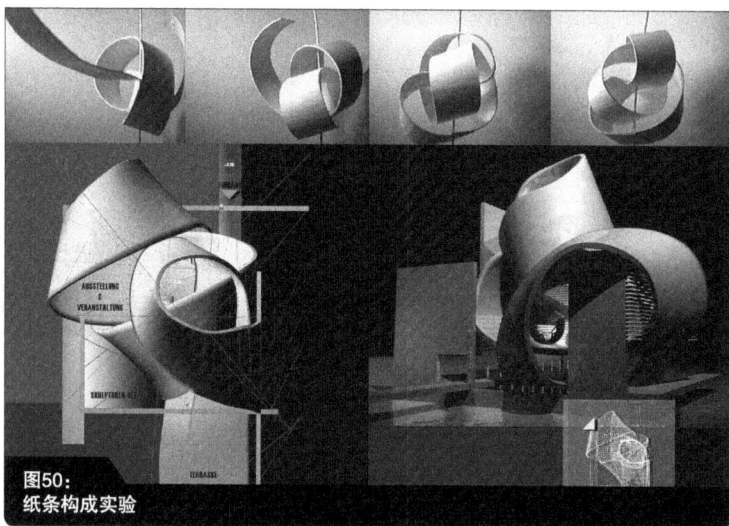

图50：
纸条构成实验

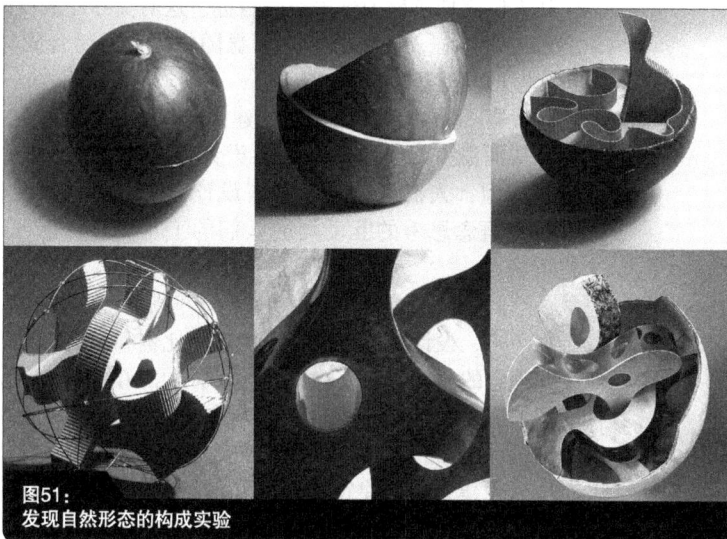

图51：
发现自然形态的构成实验

只要运用符合材料自身特殊属性（比如硬度和韧性）的方式，纸张、木料、金属等平面材料通过形态塑造都能够转变为三维形体。这些形体及其由材料属性衍生出来的形式，可以根据主要性质进行

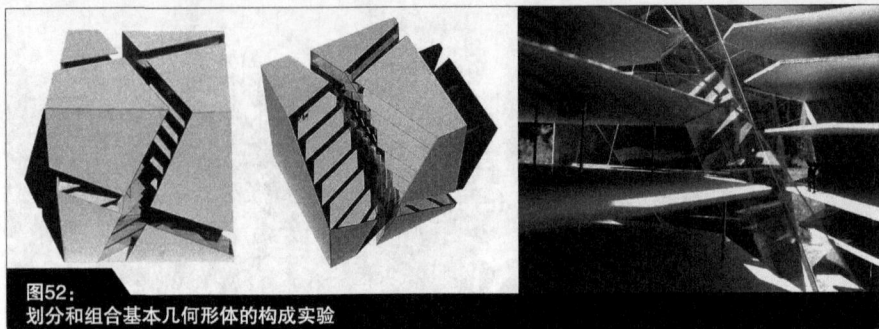

图52：
划分和组合基本几何形体的构成实验

更为精细的调节，包括张力力度、表面光线反射，以及空间塑形所带来的随机结果等等。

随着 3D 动画等技术手段的发展，采取多样的表现方式和尺度会让人们更加形象地理解空间目标。这对实际建筑设计是很实用的。通过这个过程，即便是一个明显随机的空间组合成一个建筑结构，也能够得以理解。

观察不同的物体或者空的建筑外表皮，可以引发各种各样的联想，从而激发设计灵感并引导设计。模型是理解概念的直接手段。通过模型，人们可以直接把握设计概念，并从中得到反馈。手工模型与概念思考的互动提高人们的工作信心，也可以帮助制定实施策略。

比例

空间比例表述一个空间或多个空间的长度、高度与宽度的关系。千百年来，设计必须遵循特定的空间比例和规定的几何法则与尺度。公认理想的比例关系，比如黄金分割（the Golden Section），限定了平面图、剖视图和设计细节中的所有空间要素的尺寸。一个按照黄金分割设计出来的空间会给人一种安静感和均衡感（图 53）。

纵观建筑史，建筑师们一直在尝试建立和应用标准比例的原则和系统。勒·柯布西耶模度（Le Corbusier's Modulor）试图将所有空间要素以适合人体标准比例的尺度来进行规范。他的比例原则是建立在黄金分割和斐波那契数列（Fibonacci Series）基础上的。

132

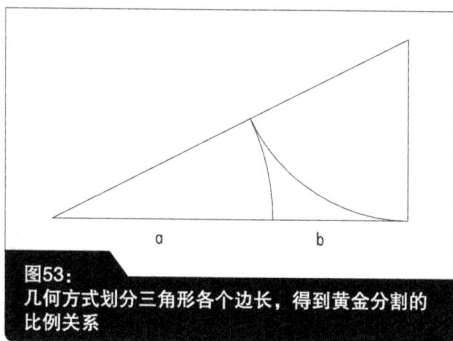

图53：
几何方式划分三角形各个边长，得到黄金分割的
比例关系

　　人们也可以预先设定结构性要素的比例关系。以砖块作为最小建筑单元，可以建立限定空间维度的模块化网格。日本的传统空间比例是按照榻榻米的数量来进行计算的。榻榻米的长宽比为 2：1（170cm×85cm），其长度是根据日本人的平均身高来设计的。一个日本传统房间由六块榻榻米组成。

　　现在最流行的十进制测量系统是基于地球的周长设计的，而不再是人体尺度。然而，在英国和美国使用的英制度量尺寸（英尺、英里等），以及一些其他传统的测量系统，仍然是基于人体尺度的。

　　正方体空间的所有边界都是等长的，会给人一种沉稳的感觉。每一个单独的空间尺寸既与其他空间尺寸有关系，也与人体尺度有关系。提高房间高度，可以增强竖向的空间效果。如果空间被拉长，

提示：
　　在黄金分割中，有两种彼此相关的长度，长边加上短边之和与长边的比例等于长边与短边的比例，即（a+b）：a=a：b。这种比例在自然和人体中都很常见。黄金分割在建筑设计、艺术、音乐等领域都被广泛应用，被认为是一种平衡与和谐的比例。它的比值是 1.618：1。

提示：
　　有些空间比例会受大体量的工业产品及它们的运输方式影响，比如船用集装箱或者木制托盘。精选的物品和器具的尺寸反过来也会影响厨房的设计，比如厨房设施决定厨房面积。

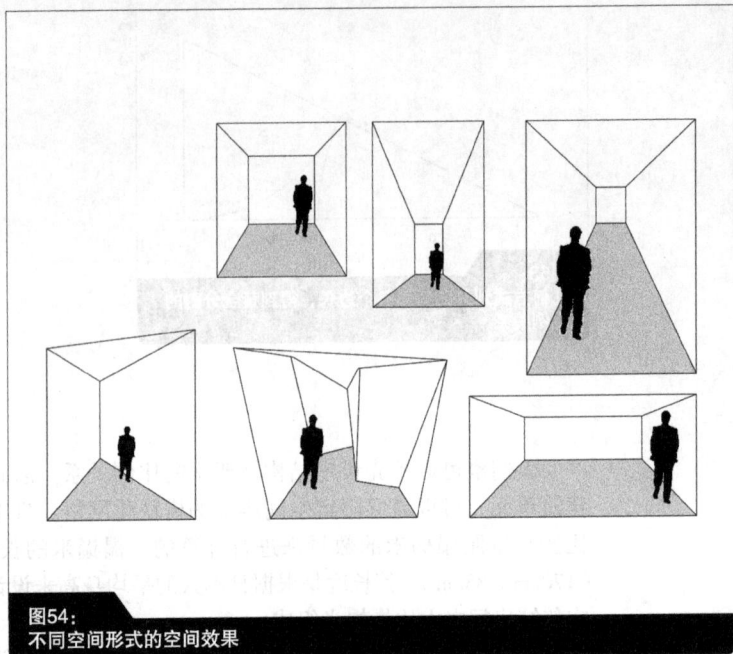

图54：
不同空间形式的空间效果

房间就会产生方向感，引导人们向某个方向移动。如果顶棚太低，甚至过低以至于人无法站立其中，那就会产生一种幽闭的感觉。在这个房间内，人们将很难移动身体，甚至根本无法移动。只能容下一人的狭窄走廊也会如此。因为这样的空间不会给人提供足够的逗留空间，人们只能寻找出口并快速通过（图54）。

维度　　　　　空间类型的差异性和通用比例的相似性是进行空间对比的一种方法。空间比例通常建立在与人体尺度或其他空间的对比之上的。紧邻超大型建筑的某个有着标准尺寸的建筑看起来会比真实尺寸小很多。反之亦然，孤零零的一栋建筑通常会显得更高一点。如果基于人们熟悉的人体尺度的要素被放置在大的房间里，那么空间的整体效果会显得更大（图55）。

空旷与密集　　　空旷和密集是空间构成的两个基本表现。人、物和相互之间可能的关联活动在空间中的容量与空间密度关系不大。我们判断空间

134

图55：
伊斯坦布尔蓝色清真寺的吊灯悬挂在顶棚的高度，通过强调宗教空间与人体尺度的高度对比关系，强化了空间的效果

是空旷的还是拥挤的，取决于人在空间中的经历，即人与空间边界之间的距离和对该空间的感受。如果感觉空间过于拥挤，人就不能在空间中自由移动，这就可能引发一种对空间的恐惧感（这也是很多恐慌发生的原因）。尽管无法被准确定义或者定量，空旷与密集之间的调节决定了观察者个人的舒适程度（图56）。

然而，因为空间密度对很多人产生相似的影响，对其进行设计可以被当作是一种空间设计的手法。增加建筑之间或物体之间的距离可以降低空间密度，加大空旷的感觉。在这种情况下，如果缺少可以测量距离的空间参照点，我们很难进行自身定位。空间是通过要素间的相互作用以及可感知的中介空间而创造出来的。在沙漠里、在无尽的黑暗里或者在无边的大海上，只能看到部分的空间边界或者压根看不到任何边界，这种空旷的感觉会让人产生恐惧感。

图56：
对城市建筑与空间进行提炼概括，可以表现出柏林和开罗两座城市不同的城市密度

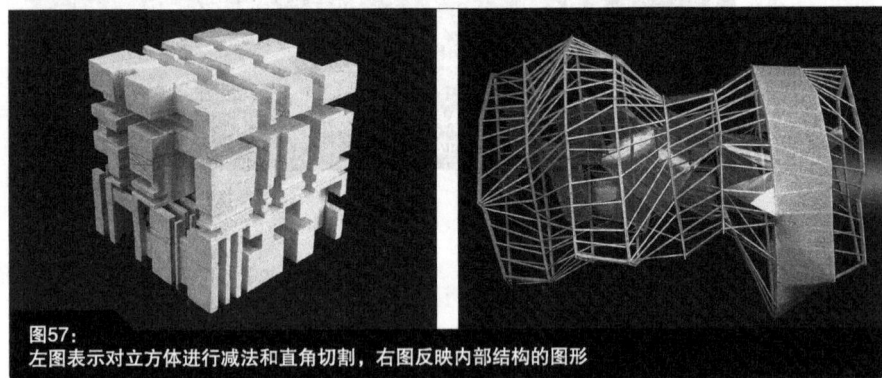

图57：
左图表示对立方体进行减法和直角切割，右图反映内部结构的图形

P60   空间、设计、结构
        各种各样的结构或形式塑造空间形态，由重力因素决定的构造结构也同样能够塑造空间形态。
        因此，我们既可以将建筑支撑结构设计成单调、坚固但看不出内部结构的形态，也可以设计成精巧、开放的形态（图57）。
        空间结构赋予建筑以独特的形态，例如平滑流动的空间、被明确划分的空间、整合在一起的闲置空间、大空间里的小隔间、直角的或自由的空间边界等等。

空间外壳

空间边界与联系

建立空间边界是空间设计的基本方法。空间边界将地球表面不计其数的空间划分成小区域。当几个空间边界共同界定空间的长度、宽度、高度时，就形成了一个空间外壳。一个独立的线形边界界定内部与外部。空间外界面可以抵御严寒酷暑、风霜雨雪、噪声以及令人不适的视线干扰。内外部空间的可渗透程度决定了空间外界面是开敞还是封闭。

开敞或者封闭的形式是由空间边界的属性及其划分方式所决定，也受空间光线条件和比例关系所决定。一些结构部件产生的阴影可以强化空间边界的效果，赋予楼板、墙面和顶棚以立体感。

空间连接

当在空间边界或者单个房间安置门窗时，就产生了内外部空间及其相互联系。这些门窗提供了与相邻空间的交通和视觉联系，并建立了水平和竖向上的关联。这时，空间外界面就像被门窗打了孔洞的薄膜一样。空间联系的数量和形式，或者界面的渗透性作为重要的设计要素决定了空间的表现形式（图58~图62）。

门窗的效果经常取决于这些门窗是空间的出入口，还是只起到联系内外视线的作用，换句话说，取决于空间边界能否穿越和如何穿越。门窗也为室内提供阳光、新鲜的空气和实现热量的交换。窗户可以建立内外双向的视觉联系。车库尺寸依据小汽车的尺寸，而门的尺寸则应依据人体尺度，吸引客人但阻挡不受欢迎的人进入，进而保障私密空间的安全性。正因为如此，门窗的开敞程度一般通过门、门帘、百叶帘、百叶窗的设置而有所差别。

提示：
    重力和荷载的传递既可以在设计中形象地强调出来，也可以对比其他设计要素进行弱化处理。

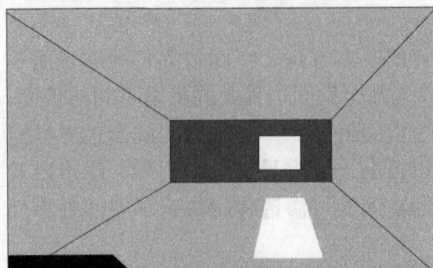

图58：
单侧直射的光源：照在墙和地板上产生强烈的光线对比，因而光线有一种耀眼的感觉

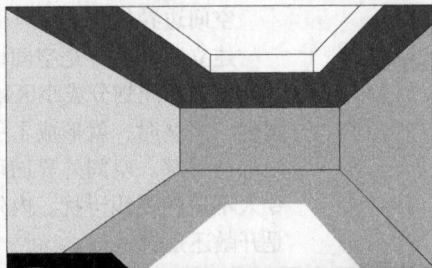

图59：
头顶的光源：漫射的光线缩小了空间的进深感

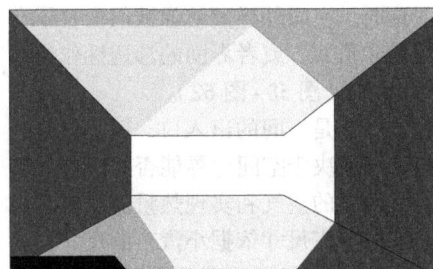

图60：
过强的光源在空间里形成了亮面和暗面

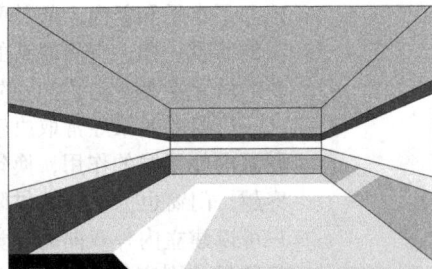

图61：
四周的光源：人们可以感受到四面环绕的光源带，较矮的墙裙增加了空间的面宽感，使空间显得更大。同时，在地板上呈现出生动的光影效果

如果墙面开口尺寸只有 70 厘米 ×200 厘米，而且还有过梁的话，空间会显得非常封闭，并且与相邻空间完全隔离。反之，如果开口宽度加倍，并且是落地窗的话，空间的边界看起来就消失了，这给人带来两个空间相互流动、融合的感觉。从地面到顶棚的自然无缝过渡可以增强这种效果。与普通窗户相比，落地窗会营造出更加开阔的空间感。与顶棚同高的带状玻璃强调了水平的方向。从另一方面来说，这种玻璃幕墙几乎完全消融了内部与外部的界限，空间向远处延伸，进而使人们暂时忘记了身处室内。空间边界的消融产生

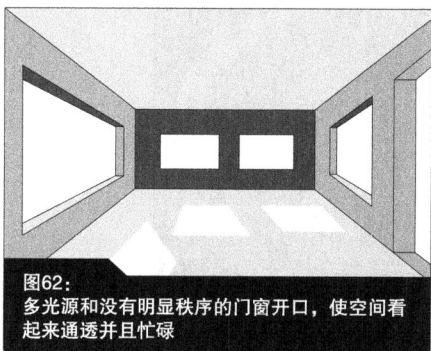

图62:
多光源和没有明显秩序的门窗开口, 使空间看起来通透并且忙碌

了新的视线或通道, 这使空间显得非常开敞, 并强调了对角方向的轴线。消融的边界降低界面的围合感, 使结构看起来缺乏稳定感, 进而可能会产生忧虑感。

门窗的位置与方向在水平与垂直剖面上限定了墙体和顶棚。墙体和顶棚上的开窗让相邻空间的景色成为这个空间的图案和要素, 门框和窗框则成为画框。

› 🖉
竖向联系

空间通过楼梯、坡道、电梯、梯子, 或者通过楼板和屋顶开口实现竖向上的联系。竖向联系是在空间进行斜向移动的必要因素, 它们可以引发移动的行为, 同时也会营造一种让人担心的感觉。依据美学原则设置的坡道和楼梯踏步, 连接某个空间或者某个特别指定的区域, 这些区域, 例如楼梯、楼梯井、电梯, 能够有效地控制

🖉

提示:

门槛表示内外部空间的转换。纵观建筑历史, 不同文化背景下设计的门槛强调内外部空间的转换, 并将空间边界特征加以表现。高门槛或者有颜色的门槛强调了内外之间墙体的空间区域。但如今越来越多建筑不需要设计门槛和栅栏, 我们也应该加以考虑。

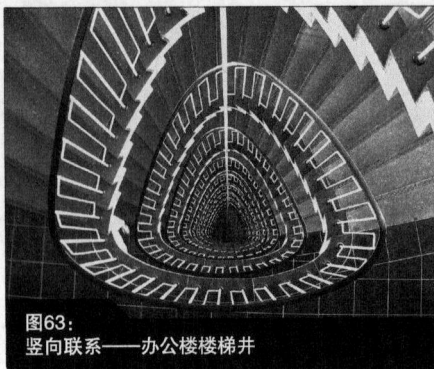

图63:
竖向联系——办公楼楼梯井

声音的传播或者抑制火灾的蔓延（图63）。楼梯间的设计强调和强化了竖向和斜向的交通及视线联系。

非常高的空间可以局部包含两层空间。如果这些双层空间面向高空间（比如美术馆）方向开敞，那么整个空间会流动起来，成为有着不同高度分区的一个空间。另一个可能的联系方式则是利用不同的标高来限定楼层，比如在楼层之间安置夹层。

P64

分层

不论材料表面或者三维实体有多大，我们都可以看到与其前后、左右相邻的其他表面。分层方式影响空间，决定空间深度效果，也会在强调变化的同时强调随之而来的差别。

🔎

示例：

一系列圆柱将长廊划分为若干个片段。这些片段更接近人体尺度，而不是整个建筑的尺度。建筑的真实长度有可能给人带来不舒适的感觉。这些圆柱创造一种韵律感，欢迎并引导人们沿着长廊通过这个空间。

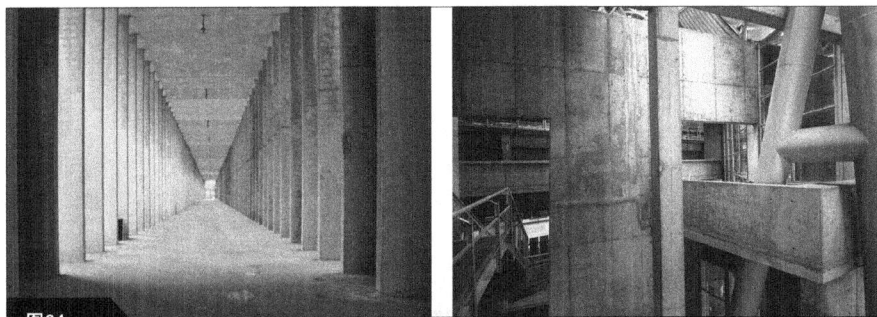

图64：
建筑构件的分层能够加强空间纵深感。不管这些分层是简单还是复杂的，人们都能够很快地获得感知

对以空间序列方式安排的要素进行定位及设计，并因而对空间深度效果产生影响。随之产生的分层建筑构件将空间进深划分为一个个的区域或片段，从而增加空间的延伸感（图 64）。

竖向分层

考虑重力因素，分层也是必要的竖向设计原则。建筑是由建筑构件一层一层叠加起来的，这些堆叠要素强调了重力和垂直轴线。如果能够看到内部堆叠或分层的要素，那么整个建筑看上去清楚明了。因此，高层建筑立面一般都会运用开窗、檐口和女儿墙来强化垂直方向上的效果。

层次感与
明晰性

在有着不同时期构成要素的空间分层上，人们可以读出空间的历史和起源。人们还可以发现不同年代的表面和结构要素。空间的特定要素能够展示历史的细节，能够激发人们对空间所在年代、空间历史及其曾经的使用者的想象。如果人们在重新设计的建筑中依然可以看到历史的痕迹，那么时间就成了有形的要素。城市之间、城市的不同区域和景观往往也是各种空间规划和随机设计叠加的结果，使城市或景观拥有了错综复杂的特征，人们往往不能一下子获悉全部信息。

P65

明晰性

人们透过不同形式和不同透光程度的门窗，可以看到空间外界面到底遮挡了什么。透明度与遮挡性也是一种用来控制公众可达性和私密性的设计方法。一面实墙完全遮挡其背后的东西，而玻璃墙

图65：
玻璃表面的反射可以降低透明感，并根据光线条件形成独特的影像

则是完全透明的（透过有光照射的窗帘，人们最多能够看到室内物品的轮廓）。

当空间的深度、属性、布置及其序列都非常清晰的时候，这个空间就有一种明晰的感觉。当人们能够明确地知道自己的所在地点和方位，并且很容易地找到出入口的时候，空间看起来也是明晰的。当人们能够看到建筑内部功能，或者是看到建筑平面图的时候，建筑也具有了明晰性（图 65）。

编排空间

空间编排是用来设计空间序列的方法。它引导着空间中人们的活动和行为。一般而言，人们只有在通过空间时，才能了解空间。他们的路线是自由的，但也取决于空间的属性及其编排顺序（图 66）。

这里有一个运用大量对比手法的空间编排的例子：在狭长的走廊尽头安排一个圆形的空间，这个空间没有明确的方向性，但有一个很高的楼梯井。这个空间编排可能会给人一种很不安的感觉，但也可能会激发人们的好奇心，进而去探索楼上的空间。狭窄低仄的教堂前厅是内外部空间转换的过渡空间。它有意地将人们引入其中，

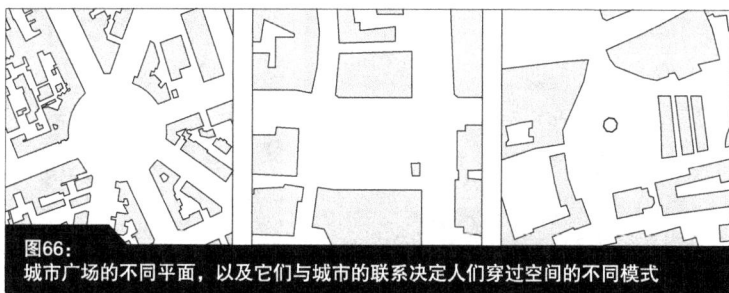

并强调出大厅内部空间的高耸感。空间比例和空间序列特性可以影响人们在空间里的行为及其路线。

在经过特殊设计的空间中，人们会采取与设计场景相协调的行为举止，就像舞台剧中的演员一样。类似地，为庆祝活动而将空间装饰一新，会使人们暂时从日常生活的情境中转换到特殊的氛围中。

空间序列设计可以富有新意，以至于它可以挑战人们的所有感官。水平和竖向穿越的运动流线可以设计为直线、折线或是曲线，进而能够影响人们行进的速度。

中世纪城市往往看起来像迷宫一样，因为他们具有丰富的空间感受，调动所有的感官并挑战人们的行动能力。这些行动流线中，只有一部分是长的线性元素，更多表现为不断变化的方向引导、多

示例：

空间可感知序列能够增加人们穿越的好奇心和动机。在两面开敞的独立房间，人们会在光线的引导下从门移动到窗。当在另一个房间中，两个敞开的门也会引导人们的行动从一个门走到另一个门。

143

图67:
充满多样性的交通区域和单调枯燥的交通区域（地铁站）

种多样的空间比例、一系列长短不一的路线。一个中世纪广场只要没有诸如汽车等潜在的危险，就更容易吸引人们在此逗留，而不仅仅是从此穿过。如今，大部分城市是依据机动车的行进尺度而进行设计的。这些城市通常都会阻碍人们步行和骑车，因而变得单调乏味（图 67）。

由于重力原因，地球对物体有向下的引力，因而垂直运动比水平运动更困难。特定垂直运动的序列、舒适度、韵律和持久度都是楼梯或坡道设计时必须考虑的因素。一般在楼层中间设计一处缓台，为人们提供休息和改变方向的空间，人们可以逐段进行攀爬。

设计者对空间中预期的运动模式也影响着空间的形式。以图形方式记录空间中最主要的重复性活动，以此作为空间形体设计最为直接和可行的基础性参数，进而确定空间的形式。因此，这些参数的方向、数量、密度和速度都起着决定性作用。建筑设计成为一种对每天不同时段预期活动的处理。人们可以从经验上预测活动的顺序，以建立与这些活动相适应的空间形式，甚至包括家具、物体、机械设备等。

P68 ## 光与影

对于人类来说，光线是电磁辐射的可见部分。在物理学上，"光"代表着整个电磁波谱。来源于太阳的自然光和来自于多种电子光源的人造光都可以被空间表面所反射。空间只有在空间各个边界和各

144

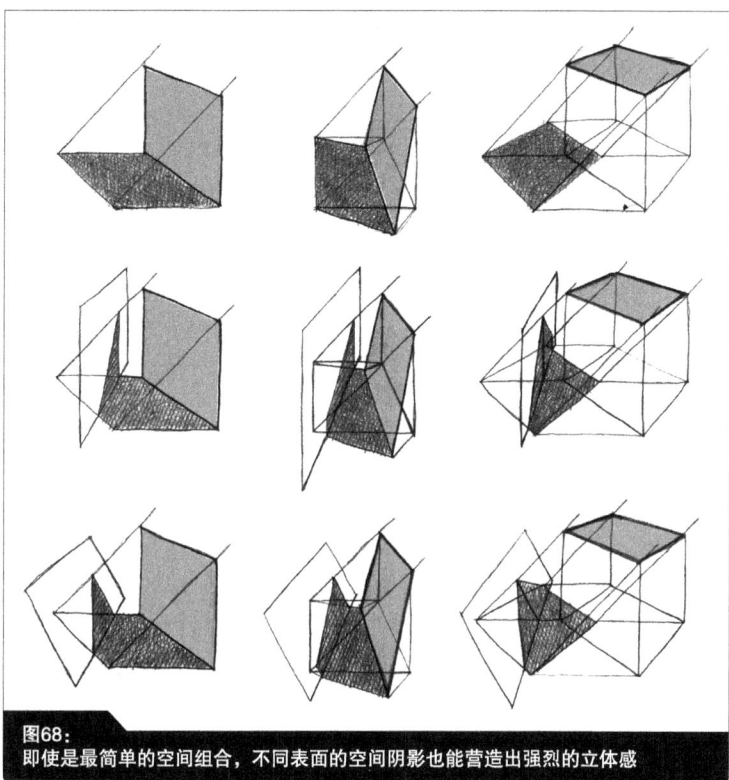

图68：
即使是最简单的空间组合，不同表面的空间阴影也能营造出强烈的立体感

个维度都充分可见的情况下才能被感知。因此，空间设计往往也是
对光线的设计。空间边界表面或强或弱地反射入射光，人们因此能
够判断出空间的范围，并能够确定自身在空间中的方位。

提示：

　　人类肉眼可见的光波波长约为380~780毫微
米（nm），与其相对应的频谱约为789~385THz。
由于人类肉眼辨别光谱的灵敏度是逐渐而不是突
然降低的，因而并没有一个确定的边界点。

紫	蓝	绿	黄	橙	红
380–420	420–490	490–575	575–585	585–650	650–750

中欧地区典型的光照度（单位：勒克斯，lx） 表5

晴天	100000
夏季多云	20000
冬季多云	3500
演播室	1000
室内与办公照明	750
走廊照明	100
街道照明	10
1 米以外的蜡烛	1
满月的夜晚	0.25

没有空间、物品、特定物质以及空气中的水分，就没有办法实现对光线的反射，光线因此就无法存在。光线可以部分渗透通过不透明的纺织物表面，也可能完全不渗透。根据光源位置、强度和方向，光线会产生阴影（图68）。

光线颜色

> ⌇ > ⌇

光线在人类可见的范围内划分为几种光谱颜色。相应的光谱颜色取决于连续光谱的最大波长及其对应的颜色温度（TCP），以热力学温标开尔文（Kelvin，K）进行测量（表4）。

⌇

提示：
灯泡的灯丝有一个相对稳定的波长，因而可以产生和太阳光相似的连续光谱。

⌇

提示：
与噪声或废气一样，光线也是一种环境因素。从光源装置散发出来的光线能对人类和动物产生伤害，甚至能够对机械产生破坏。

图69:
头顶和侧面的光源可以增强空间的立体感

光线和颜色的度量方法很相似，对光线的感知因人而异，并且很难达成共识（表5）。

明与暗是空间设计的必要元素，受到光源选择和空间结构表面的影响。光线在每年的不同日期和每天的不同时刻都会变化，因此是一个不稳定的设计要素。作为空间设计的方法，光线应根据不同的功能和不同的时间加以设计。空间可以依据功能和具体要求来引入或屏蔽光线。不同的进光类型和位置会产生不同的效果。

提示：

坎德拉（Candela，为拉丁语，相当于一只普通蜡烛的发光强度）：发光照度单位，用符号"cd"表示，是流明（lumen）的基本测光单位；流明（Lumen，为拉丁语，意为光线，灯）：光通量的单位，用符号"lm"表示；勒克斯（Lux）：发光照度和光发射度，用符号"lx"表示，可以用照度计进行测度，能够换算为光通量和光照度。

温度、湿度、声音、气味

           空间的物理状态也是空间设计的工具。但是因为物理状态是不
      断变化的，需要一个更加动态的方法。空间状态取决于空间外表面
      的材料性质，或者内外部之间的媒介物。

温度   材料的导热性能决定了从人体或空气中快速传递热量和内外部
      温差平衡的方式。材料的导热性与其表面温度有着很大的不同吗？
      尽管有充足的供暖，但不合时宜的空气流动会使人们感觉很冷吗？
      地板、门把手、座椅等人们经常接触的建筑部件的导热性能也应该
      加以考虑，这样才能营造一个舒适、宜人、愿意在此停留的空间（比
      如赤脚走进起居室）。

           房间温度是根据空间特定功能而定的。比如体力劳动者比脑力
      劳动者需要更低的温度。老年人通常比年轻人需要更高的温度。对
      冷和热的感知也是因人而异的。室内供暖系统可以是皮肤能够接触
      到的直接传热型（例如太阳光和暖气管），也可以是空气传导的对流
      传热型。房间温度对房间的湿度也有影响。

湿度   空间表面及其温度会吸收水蒸气。湿度是用来表示一个空间内
      或在地面上的混合空气里水蒸气所占百分比的物理量。湿度的百分
      度表示空气中水蒸气的饱和程度。

           几乎在进到房间的一瞬间，皮肤就可以感受出湿度。房间的相
      对湿度可以影响人体的舒适度，进而影响人的健康。比如说，湿度
      高会使灰尘聚集在一起，而湿度低会使鼻子发干，都可能引发疾病。
      空间中潮湿的表面容易滋生细菌，也会对人体健康产生危害。

示例：
    触摸石头与触摸木材相比，手掌温度更快
地传递给石头。因为石头可以让手掌更快地降
温，所以即便两者表面温度是一样的，石头摸
上去也比木材更凉一些。

提示：
    相对湿度 50% 意味着空气中只含有水蒸气
最大容量的一半，相对湿度 100% 则意味着空气
中水蒸气含量已经完全饱和。如果超过了这个含
量，多余的水分就会形成凝结聚集在空间表面或
者变成雾。

不同功能空间的混响时间（频率范围为 100~5000Hz）			表 6
录音棚	教室、报告厅	混合办公室	音乐厅（根据演奏的不同音乐类型决定）
< 0.3 秒	0.6~0.8 秒	0.35 秒	1.5~3.0 秒

声音 可以通过声学方面的空间设计控制干扰性的噪声。噪声影响空间的音响效果，并且很快就会被察觉到。空间的外界面可以很好地保护空间内部不受外界噪声干扰，反之亦然。材料的声学效果取决于对声音的吸收能力，数值介于 0 和 1 之间，0 是完全不吸收，1 是完全吸收。声音的吸收程度依赖于碰撞频率。以下两种不同音响类型有着明显的区别：多孔的消声设备可以将声音吸收进入材料。在孔隙内部，摩擦将声音能量转化成热量，因此减少了材料对声音的反射；依靠振动的消声装置则因声波冲击而产生震动，这种共鸣也降低了对声音的反射水平。

空间音响效果的最重要因素是混响时间。混响时间指的是一个声音在空间中的延迟时间，需要根据空间的功能加以设计（表 6）。一个音乐厅内部的声学设计应该尽可能精细，同样的，大型办公室的空间也有着具体的声音要求。大型宗教空间一般都有很长的混响时间与强烈的声音反射，相比之下，非常封闭的空间则显得小而幽闭。

气味 材料的挥发和其他人的体味都会影响人们对所在空间氛围的感知，例如在图书馆、教堂、学校、储物室等空间内。强烈的气味可以压倒空间的所有其他因素，进而使空间发生变化。香味如果能够建立正面的联系，就能营造一种愉快的空间效果。购物商场和百货商店的特定区域经常运用香味给人们带来愉悦感。

P74 **材料、质地、装饰、色彩**

组成空间的各个表面对人们的影响取决于材料使用的所有方面。除了具体的材料质量或者材料表面的质地以外，材料的基本性质也会影响人们的空间感受（表 7）。

材料质感首先取决于材料的手工或机械制作的方式，但也是使用、老化或腐蚀的结果。大多数材料的表面质感可以被描述为粗糙的、精细的、光滑的、暗淡的，有光泽的，等等。

表面质感也能影响光环境、声环境、温度以及房间湿度。

如果空间界面能够随着人们的活动而变化，那么这个空间就会像一件贴身衣服一样，实用且舒适。

新型材料

纳米材料和合成材料等新型材料不断被研发出来，并用于空间设计。纳米技术已经使新型材料表面、涂层以及质感能够满足特定的需求。由于表面设计的改变非常微小，所以人类肉眼根本无法看到。

"合成"指的是使用新研发的材料混合物，以此来提高建造质量。例如，研发新矿物骨料时发明了透明的水泥。某种升级材料用合成材料取代了钢筋，就很有可能建造出更加轻薄的楼板。

纤维和纸张

因为纤维和纸张很容易在使用过程中被损坏，只能作为临时的空间设计元素。和家具一样，纤维和针织品可以随处移动和使用。它们非常柔软、轻便，有着多种多样的纹理和丰富的色彩，可以用来表达设计概念。因此，它们被广泛地运用到墙体、顶棚、地板、家具等的设计中。空间设计中典型的纺织元素和材料包括：

- 地毯（地面或墙面）：羊毛、合成材料、棉花、丝绸、黄麻、剑麻
- 墙纸 / 挂毯（墙面和顶棚）：羊毛、棉花、丝绸、亚麻、金属
- 窗帘（内墙和外墙）：棉花、羊毛、合成材料、丝绸、亚麻、金属丝
- 毛毯与枕头：（家具）：羊毛、棉花、丝绸、亚麻、合成材料

相比于石头和金属，纺织品和纸张的特性与衣服更为相似，因此，人们对其感觉是舒适且熟悉的。因为它们大多柔软、富有弹性且不产生热量传递，因而人们更愿意触摸它们。

提醒：

　　用于空间设计的多种材料能够表现为一种材料拼贴。这种拼贴显示项目所选取的不同材料之间的相关联系。

提示：

　　一个人沿着木质地板行走的方式在不同的铺设情况下是截然不同的，比如在钢筋混凝土建筑中砂浆铺设，或在薄钢板上铺设。这些差异源自材料的弹性、声音以及热量吸收等特性。

表 7

空间设计可用的材料（摘选）

自然石材	人工石材	木材	玻璃	天然纤维和织物	金属	合成材料和复合材料	其他
深成岩和喷出岩 – 辉长岩 – 花岗岩 – 闪长岩 – 玄武岩 – 辉绿岩 – 浮石 – 玄武岩熔岩 – 斑岩 – 石灰华 水成岩 – 板岩 – 石灰石 –（石灰沙砖 – 盆缘石灰石） – 砾岩 – 白云石 – 杂砂岩 – 砂岩 – 石英岩 变质岩 – 片麻岩 – 大理石 – 石英岩 – 板岩	砖头 硅石砖 渣块 陶瓷板（棉） 土砖 铸石 / 混凝土（水泥、水、沙 / 砂石） – 混凝土砌块 – 水磨石 – 预拌混凝土 – 混凝土块 – 砂浆 –agglo 大理石 – 透明混凝土 – 水泥石灰石 陶瓷 石英材料	软木 硬木（包括果木、树瘤木） 热带木材 竹子 麦草 芦苇 所有木质材料 木纤维 软木 椰壳	工业玻璃 夹层安全玻璃（LSG） 玻璃窗 玻璃管 夹丝玻璃 玻璃纤维 玻璃石 玻璃纤维 玻璃马赛克 玻璃珠	– 毛毡和皮毛 羊毛 – 针织品 – 藤条 动物纤维 （纺织或条播） 毛（羊毛、羊驼毛、美洲驼、安哥拉山羊、羊绒、骆驼毛、马海毛） 毛发（山羊、奶牛 /牦牛、马） 丝（桑蚕、柞蚕、贝壳） 角质 皮毛 皮革 植物纤维 – 麻布 – 苎麻 – 亚麻 – 椰子 – 棉线 – 木棉 – 大荨麻 – 大麻 – 黄麻 – 剑麻 – 竹子 – 草 –（草壁纸）	铸造金属、轧制金属 合金 铁（钢，不锈钢） 铜 铅 镍 铝 锌 锡 钛 银 金 固体 （管、杆） 板材 管线 纺织品 薄膜 薄板 合金 泡沫	麦草 土坯 油毡 沥青 环氧树脂 有机玻璃 亚克力 泡沫橡胶 矿岩棉板 薄膜薄片树脂（涂层） 热塑性塑料 – 聚乙烯 – 尼龙 –PET 塑料 – 聚苯乙烯 – 聚酰胺 – 聚酯 – 聚丙烯 热固性材料 – 聚酯 – 酚醛塑料 – 尼龙 – 聚亚安脂 – 合成树脂 – 环氧树脂 – 三聚氰胺 弹性材料 – 塑胶（橡胶） – 聚亚安酯	矿石 无机粘结胶泥 白灰石膏 洋灰石膏 石膏灰泥 新材料 纳米材料（1–100 纳米） 油漆涂料（丙烯、合成树脂） 纤维材料 微纤维 矿物纤维 石棉 自然聚合纤维 – 黏胶 – 锦纶（尼龙） – 莫代尔 – 卡纸板 – 纸 – 壁纸 – 纸型 化学和陶瓷纤维

因为纺织品和纤维制品的多孔结构，并且具有较大的表面积，在空间设计中常被用作吸声材料。它们还可以用来表现空气的流动，并且能够用来遮风挡雨。

装饰

装饰一词用于说明抽象或具象的形式或物体的反复使用（拉丁文中 ornare 即为装饰）。创造的形式可以用来设计和构造空间要素与界面。装饰也代表某种象征意义，有一种"题字"的效果。墙纸一般都会有装饰性的图案，但是天然石头引人注目的纹理式样也同样具有装饰性。在城市平面图中，重复的建筑排列能够创造一种形式。可以对不同人行道铺装相连的节点进行装饰性设计（图 70）。

许多抽象的装饰图案是源于自然形态和图像的。装饰物可以绘制在一个表面上，也可以作为立体要素来构建房间。哥特式教堂的柱子和墙上有许多装饰性的结构元素以及装饰物，通过光影的反射来增强建筑的立体感。

装饰经常被用来组成较大的表面和空间，使得它们不至于显得过于巨大、空旷甚至恐怖。基于人体尺度对大的表面和空间进行划分，使得人们能够很容易地明确自己在空间中的定位。装饰会让一个空间充满了视觉和触觉的感官刺激，这样就能减少或者消除空旷的感觉。然而，过于多样化的结构或者过度装饰的立体装饰品反而会破坏空间感。

色

空间中的每一个结构表面都可以通过反射自然光或者人造光获取色彩感知。色彩效果取决于材料及其质感和表面特性。空间的整体色彩效果具有不同的强度，这取决于光线、反射的程度、角度以及材料表面的色彩属性。

提示：
　　由于禁止使用图像，伊斯兰教徒以文字为基础逐步形成非常精细、复杂和多样化的装饰物，并将其用在织物和空间设计上。

图70:
公共空间中的装饰

人们常利用色彩来对空间进行分区，对房间的焦点加以强调或淡化。在有着黑色地板的白色房间内的顶棚看起来比在有着白色地板的黑色房间内要高。色彩还能够影响我们对空间边界的感知：浅色的、低对比度的色彩相比深色的、高对比度的色彩能让空间显得更宽更大，这是因为在这样的空间中，人们不容易判断出与各个界面之间的距离。

如果采用冷色调、低对比度的颜色作为背景，空间的纵深感将会得以提升。

P78

家具：固定和可移动的元素

通常来说，在空间各个表面之间能够创造空间的一切事物都是空间设计要素，既包括固定的要素，也包括可移动的要素。柱子是永久固定的支撑元素，但家具却是可以自由移动的。塑造空间体量的各种要素之间的互动关系总是对空间产生影响。即使其中某个要素相比其他要素具有更大的影响力，但整体空间质量仍然是所有要素效果整合的结果。这些要素之间相互关联，并塑造要素之间的空间（图 71）。

家具

家具是一种空间设计要素，既可以永久固定在墙上，也可以自由移动。它在基本的空间界面内部形成了灵活的次级结构，是实现空间个性化的重要工具（图 72）。

家具之间形成了各自的联系，一个桌子会搭配不同的椅子，一组家具在室内创造出一片小区域。家具可以让一个原本熟悉的空间变得很陌生。然而从另一个角度来说，家具可以支持甚至提升现有的空间价值和空间初始结构的效果。

提示：

对色彩的感知因人而异，很难用语言来描述色彩对空间的影响。因此，色卡和色标可以用来帮助人们对色彩进行选择并加以表达。

提示：

色彩甚至可以影响人们在空间中停留的时间。经验表明在强烈的橙、黄和红三种配色中的停留时间少于在绿和白的配色中的停留时间。因此，快餐店一般会选择橙、黄和红的配色方案，顾客在进餐后会尽快离开，进而为下一位顾客出空间。

家具的尺寸鼓励人们对其触摸和使用。因为家具是基于人体尺度进行设计的，并能够在日常空间体验中对空间进行二次划分，家具成为非常常见的设计要素。

　　在公共空间、公寓或办公室里，因为家具可以对空间进行构建，并赋予不同的用途，它经常会被用作一个可变的空间要素。家具也是能够减小或者快速有效地增加空间密度的工具。家具帮助人们建立与空间的身体联系。房间里的一把椅子具有吸引力，只是因为它在暗示人们过来坐下。

　　在私人房间内，总有一些家具是与家庭的记忆或者特殊事件相关联的。因而，这些家具的形式和特性变得不那么重要，而成了个人感情和记忆的载体。

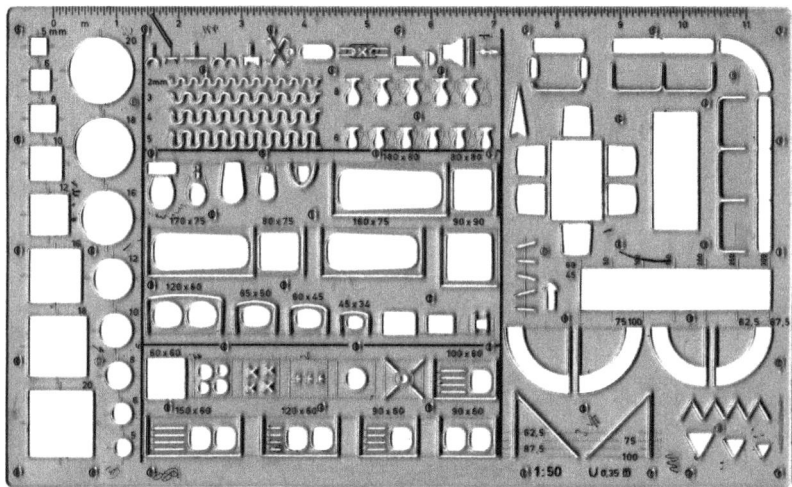

图71:
家具模板

图72：
可调节的家具要素能够适应既定的空间形式。但右图所示的一个实用的家具则是为这个小型空间特别定制的

示例：
　　家具的尺寸区别很大：典型的四角小床能够与房间一样大，因而成为空间中的一处舒适空间。

　　　　结论

　　空间设计的目标是给建成空间或场地赋予一种态度，这种态度是能够被感官和认知系统所感知到的。这是通过满足功能、技术、知识和美学需求而实现的。

　　人类存在于时空之中。在这个过程中，人们以主动或被动的方式不断地感觉和认知多种空间特征。同时，人们采取多种多样的方法，主动或被动地改变着他们的空间环境。因此，人和空间始终处于一种持续的动态关系之中。空间设计考虑人类生存的诸多方面，能够使人们感到满足和舒适，并能够提供交流的多种可能性。

　　空间设计塑造具有特定环境氛围的建筑环境。这种氛围对空间的个人满意度、社会交往以及空间使用、活动的方式产生积极的影响，甚至就是为了促进其与空间环境之间的联系。情绪与空间氛围是各种各样的从感觉和认知出发的空间感知现象的集合。

　　空间设计是否会随着时间的流逝得到人们的认可，对其进行利用甚至进一步改善，这取决于能否成功地协调人与空间之间复杂的互动关系。空间氛围可以被设计成持久的或者暂时、随时变化的。即便在不对原有建筑结构和材料元素进行重新设计的情况下，原有空间的简单利用都能够改变空间氛围，因为已经被新的使用者重新定义。因此，空间环境氛围不可能完全地预先设计。即使新建建筑具有较为长久的使用时限，但人们对它的空间印象也不是简单地保持不变。

　　然而，空间环境氛围还是可以被设计的，这就是为什么这本书对空间设计的各种元素和方法进行了总体介绍。空间环境氛围的类型和它们的表现形式取决于诸多要素之间持续、复杂的互动，这些要素包括设计理念和空间概念、行为活动、空间使用者、空间形式、序列和肌理、空间结构的材料特征、空间的光线和音响效果等等。策略性地采用所有这些要素可以实现完善的空间设计。

　　空间设计的一项有趣任务是塑造空间的未来效果，也就是要对以下问题进行考虑并做出决策：能否和如何引导人们清楚地理解空间结构？能否和如何为人们进行个性化空间塑造提供机会？

P82　　　　致谢

感谢多特蒙德的伯特·比勒费尔德（Bert Bielefeld）耐心的编辑工作；

感谢锡根的蒂娜·杰克（Tina Jacke）的图片与表格编辑工作；

感谢锡根的佩特拉·克莱因（Petra Klein）的组织协调工作；

感谢希格露恩·姆萨（Sigrun Musa）的图片与表格编辑工作；

感谢柏林的朱迪思·劳姆（Judith Raum）的编辑工作。

P82　　　　**参考文献**

Rudolf Arnheim: *The Dynamics of Architectural Form*, University of California Press, Berkeley and Los Angeles 1977

Gaston Bachelard: *The Poetics of Space*, Orion, New York 1964

Franz Xaver Baier: *Der Raum*, Walther König, Cologne 1996

Gernot Böhme: *Atmosphäre*, Suhrkamp, Frankfurt am Main 1995

Otto Friedrich Bollnow: *Human Space*, Princeton Architectural Press, New York 2008

Michel de Certeau: *The Practice of Everyday Life*, University of California Press, Berkeley 1988

Fred Fischer: *Der animale Weg*, Artemis, Zurich 1972

Kenneth Frampton, Harry Francis Mallgrave: *Studies in Tectonic Culture*, MIT Press, Cambridge 2001

Walter Gölz, *Dasein und Raum*, Max Niemeyer Verlag, Tübingen 1970

Max Jammer: *Concepts of Space. The History of Theories of Space in Physics*, Harvard University Press, Cambridge 1954

Hugo Kükelhaus: *Unmenschliche Architektur*, Gaia, Cologne 1973

Wolfgang Meisenheimer: *Choreography of the Architectural Space. The Disappearance of Space in Time*, Dongnyok/Walther König, Paju/ Cologne 2007

László Moholoy-Nagy: *The New Vision. Fundamentals of Design, Painting, Sculpture, Architecture*, Faber, London 1939

Paul von Naredi-Rainer: *Architektur und Harmonie. Zahl, Maß und Proportion in der abendländischen Baukunst*, DuMont, Cologne 1982

Christian Norberg-Schulz: *Genius Loci. Towards a Phenomenology of Architecture*, Rizzoli, New York 1980

Colin Rowe, Robert Slutzky: *Transparency*, Birkhäuser, Basel 1997

Bernard Rudofsky: *Architecture Without Architects. A Short Introduction to Non-Pedigreed Architecture*, University of New Mexico Press, Albuquerque 1987

　　　　　图片鸣谢

Fig. 51   Kerstin Kaiser: spatial design study research, University of Siegen
Fig. 52   Mathias Both: spatial design study research, University of Siegen
Fig. 57 left   Marko Hassel: study research, University of Siegen
Fig. 57 right Maike Niederprüm: study research, University of Siegen
Fig. 63   Oberfinanzdirektion Frankfurt am Main, photo: Sigrun Musa
Fig. 64 left   Aldo Rossi: San Cataldo cemetery, photo: Bert Bielefeld
Fig. 64 right Enric Miralles and Carme Pinós: Center for Rhythmic Gymnastics, Alicante, photo: Ulrich Exner
Fig. 67 right Subway station, Seoul, photo: Stefany Kim
Fig. 69   Christoph Ahlers
Fig. 70   Stefan Schilling: Zeil Projekt, Frankfurt
Fig. 72 left   bfa, büro für architektur: apartment building, photo: Valentin Wormbs
Fig. 72 right INDEX Architekten: Die Bank—adaptable furniture, photo: Stefan Schilling

Images on the following pages are credited to the authors:
2, 3, 6, 8, 9 left, 11 left, 14–18, 21 left and right, 24 right, 26 left and right, 28–30, 32, 33, 35–39, 41, 43, 44 left and right, 45 left and right, 46–49, 53–55, 56 left and right, 58–62, 64 right, 65, 66, 67 left, 68.

P84   作者简介

欧利奇·埃克斯纳（Ulrich Exner），建筑师与工程师，锡根大学建筑与城市规划系空间设计与规划教授，美因河畔法兰克福自由建筑师。

迪特里希·普雷塞尔（Dietrich Pressel），建筑师，锡根大学建筑与城市规划系空间设计与规划助理研究员，美因河畔法兰克福自由建筑师。

译者简介
董慰，哈尔滨工业大学建筑学院副教授，博士
张宇，哈尔滨工业大学建筑学院讲师，博士